Books are to be returned on or before
the last date below.

FAILURE ANALYSIS OF INTEGRATED CIRCUITS

TOOLS AND TECHNIQUES

FAILURE ANALYSIS OF INTEGRATED CIRCUITS

TOOLS AND TECHNIQUES

edited by

Lawrence C. Wagner, Ph.D.

Texas Instruments Incorporated

KLUWER ACADEMIC PBULISHERS
Boston / Dordrecht / London

Distributors for North, Central and South America:
Kluwer Academic Publishers
101 Philip Drive
Assinippi Park
Norwell, Massachusetts 02061 USA
Telephone (781) 871-6600
Fax (781) 871-6528
E-Mail <kluwer@wkap.com>

Distributors for all other countries:
Kluwer Academic Publishers Group
Post Office Box 322
3300 AH Dordrecht, THE NETHERLANDS
Telephone 31 78 6576 000
Fax 31 78 6576 254
E-Mail <services@wkap.nl>

 Electronic Services <http://www.wkap.nl>

Library of Congress Cataloging-in-Publication

Failure analysis of integrated circuits : tools and techniques / edited by
Lawrence C. Wagner.
    p. cm. – (Kluwer international series in engineering and computer science ; SECS 494)
    Includes bibliographical references and index.
    ISBN 0-412-14561-8
    1.   Semiconductors—Failures. 2. Integrated circuits—Testing.
    3. Integrated circuits—Reliability.   I. Wagner, Lawrence C.   II. Series
    TK7871.852.F35   1999
    621.3815--dc21   98-51769
                                                                               CIP

Copyright © 1999 by Kluwer Academic Publishers. Third Printing 2003.

This printing is a digital duplication of the original edition.

All rights reserved. No part of this publication may be reproduced, stored in a retrieval system or transmitted in any form or by any means, mechanical, photo-copying, recording, or otherwise, without the prior written permission of the publisher, with the exception of any material supplied specifically for the purpose of being entered and executed on a computer system, for exclusive use by the purchaser of the work.

Printed on acid-free paper.

Printed in Great Britain by IBT Global, London

TABLE OF CONTENTS

Preface ....

Acknowledgements ....

1 Introduction
Lawrence C. Wagner 1

    1.1 Electrical Characterization   3
    1.2 Die Exposure   3
    1.3 Fail Site Isolation   4
    1.4 Package Analysis   4
    1.5 Physical and Chemical Analysis   4
    1.6 Diagnostic Activities   5
    1.7 Root Cause and Corrective Action   10
    1.8 Conclusion   10

2 Electrical Characterization
Steven Frank, Wilson Tan and John F. West 13

    2.1 Electrical Characterization   14
    2.2 Curve Tracing   18
    2.3 Electrical Characterization of State Dependent Logic Failures   25
    2.4 Memory Functional Failures   31
    2.5 Challenges of Analog Circuit Fault Isolation and Analog Building Blocks   34
    2.6 Future Challenges for Circuit Characterization   39

3 Package Analysis: SAM and X-ray
Thomas M. Moore and Cheryl Hartfield 43

    3.1 The Scanning Acoustic Microscope   44
    3.2 The Real –Time X-Ray Inspection System   47
    3.3 Application Examples   49
    3.4 Summary and Trends in Nondestructive Inspection   54

4 Die Exposure
Phuc D. Ngo..59

    4.1 Didding Cavity Packages   59
    4.2 Decapsulation of Plastic Packages   61
    4.3 Alternative Decapsulation Methods   62
    4.4 Backside Preparation for Characterization and Analysis   65
    4.5 Future Requirements   66

5 Global Failure Site Isolation: Thermal Techniques
Daniel L. Barton 67

    5.1 Blackbody Radiation and Infrared Thermography   67
    5.2 Liquid Crystals   70
    5.3 Fluorescent Microthermal Imaging   76
    5.4 Conclusion   84

6 Failure Site Isolation: Photon Emission Microscopy Optical/Electron Beam Techniques
Edward I. Cole and Daniel L. Barton 87

    6.1 Photon Emission Microscopy   88
    6.2 Active Photon Probing   95
    6.3 Active Electron Beam Probing   102
    6.4 Future Developments for Photon and Electron Based Failure Analysis   110

7 Probing Technology for IC Diagnosis
Christopher G. Talbot..113

    7.1 Probing Applications and Key Probing Technologies   113

7.2 Mechanical Probing	114
7.3 E-beam Probing	117
7.4 Navigation and Stage Requirements	128
7.5 FIB for Probing and Prototype Repair	131
7.6 Backside Probing for Flip Chip IC	137

8 Deprocessing
Daniel Yim 145

8.1 IC Wafer Fabrication	145
8.2 Deprocessing Methods	148
8.3 New Challenges	156

9 Cross-section Analysis
Tim Haddock and Scott Boddicker 159

9.1 Packaged Device Sectioning Techniques	159
9.2 Wafer Cleaving	162
9.3 Die Polishing Techniques	164
9.4 Cross Section Decoration: Staining	165
9.5 Focused Ion Beam (FIB) Techniques	166
9.6 Sectioning Techniques for TEM Imaging	168
9.7 Future Issues	173

10 Inspection Techniques
Lawrence C. Wagner 175

10.1 Microscopy	175
10.2 Optical Microscopy	177
10.3 Scanning Electron Microscopy	183
10.4 Focused Ion Beam Imaging	188
10.5 Transmission Electron Microscopy	189
10.6 Scanning Probe Microscopy	189
10.7 Future Considerations	192

11 Chemical Analysis
Lawrence C. Wagner 195

    11.1 Incident Radiation Sources   195
    11.2 Radiation-Sample Interaction   197
    11.3 Radiation Flux   197
    11.4 Detectors   198
    11.5 Common Analysis Techniques   198
    11.6 Microspot FTIR   199
    11.7 Other Techniques   201
    11.8 Conclusion   203

12 Energy Dispersive Spectroscopy
Phuc D. Ngo 205

    12.1 Characteristic X-Ray Process and Detection   205
    12.2 Quantitative Analysis   209
    12.3 Sample Considerations   210
    12.4 EDS Applications   212
    12.5 Future Considerations   214

13 Auger Electron Spectroscopy
Robert K. Lowry 217

    13.1 The Auger Electron Process   217
    13.2 AES Instrumentation and Characteristics   219
    13.3 AES Data Collection and Analysis   220
    13.4 Specimen, Material, and AES Operational Concerns   221
    13.5 AES in Failure Analysis   223
    13.6 Conclusion   226

14 Secondary Ion Mass Spectrometry, SIMS
Keenan Evans 229

    14.1 Basic SIMS Theory and Instrumentation   230
    14.2 Operational Modes, Artifacts, and Quanitification   233

14.3	Magnetic Sector SIMS Applications	234
14.4	Quadrupole SIMS Applications	236
14.5	Time-of-Flight SIMS Applications	238
14.6	Future SIMS Issues	239

15 Failure Analysis Future Requirements
David P. Vallett 241

15.1	IC Technology Trends Driving Failure Analysis	242
15.2	Global Failure Site Isolation	243
15.3	Development in Probing	245
15.4	Deprocessing Difficulties	246
15.5	Defect Inspection – A Time vs. Resolution Tradeoff	246
15.6	Failure Analysis Alternatives	248
15.7	Beyond the Roadmap	249

Index 251

PREFACE

This book is intended as guide for those diagnosing problems in integrated circuits. The process of selecting the tools and techniques to apply to a specific problem can be bewildering and inefficient. As shown throughout the book, there are a wide variety of tools and techniques employed in the diagnosis or failure analysis of semiconductors. It is not practical to attempt to apply all of them to any specific failure. Some are mutually exclusive. For example, it is not practical to attempt to perform TEM and deprocess the exact same device. Some techniques are targeted as specific failure mechanisms. Hence, it is not practical to apply them unless there is an indication that the particular failure mechanism is possible. Further, the cost and effort to apply all of the available techniques make it impractical.

This book provides a basic understanding of how each technique works. It is, however, not intended to provide mathematical detail. Rather, it provides the qualitative understanding generally required in making intelligent tool choices. The book discusses the shortcomings and limitations of each technique as well as their strengths. Typical applications of the techniques are used to illustrate the strengths of the techniques. The book is also not intended to provide recipes for executing those techniques. Those recipes are very dependent on the semiconductor manufacturing process and the specific tool manufacturer. The diversity of semiconductor processes and tools make it impractical to attempt that in a single volume.

As well as understanding how and when to apply each technique, it is important to understand how they fit together. It is all together too easy to become tied up in attempting to make one technique work on a specific failure. The diagnosis is really part of a much bigger continuous improvement process. That process and the integration of the tools are presented in the first chapter.

The semiconductor industry is propelled by the rapid pace of technological improvements. The trends towards more complex, faster, denser devices with smaller features and more layers. None of these trends makes diagnosis of problems easier. Thus the technology of failure analysis must strive to keep pace. The efforts of failure analysis to keep pace are described in the final chapter.

Acknowledgements

I would like to thank all of the chapter authors who contributed so much of their time and effort to making this book possible. I would also like to thank the people who helped with chapter reading including Richard Clark of Intel, Dave Vallett of IBM, Ken Butler, Hal Edwards, Walter Duncan, Gordon Pollack, Tim Haddock, John West, John Gertas, and Monte Douglas from Texas Instruments.

I would to like to especially thank Randy Harris for his support in completing this book.

1

INTRODUCTION

Lawrence C. Wagner
Texas Instruments Incorporated

Failure analysis can be defined as a diagnostic process for determining the cause of a failure. It has broad applications in various industries but is particularly important in the integrated circuit industry. Within the semiconductor industry, failure analysis is a term broadly applied to a number of diagnostic activities. These activities are geared towards to supporting the determination of a "root cause" of failure, which support process improvements impacting product yield, quality and reliability. In its most narrow usage, failure analysis is the analysis of completed semiconductor devices, which have failed. These consist primarily of customer returns, qualification failures and engineering evaluations. In the broadest sense, failure analysis includes a diverse range of diagnostic activities, which include support of process development, design debug, and yield improvement activities in the both the wafer fab and assembly site. Despite the broad range of diagnostic activities, a relatively common set of tools and techniques has emerged to support them. However, nearly all of the extremely diverse group of the tools and techniques are used in the failure analysis of completed devices. Because of this, the failure analysis of completed devices provides an excellent perspective to approach a discussion of these tools and techniques.

These tools and techniques can be grouped into several critical subprocesses, which form a generic failure analysis process flow. This is illustrated in Figure 1.1 as a flow chart. This flow for failure analysis of completed devices includes all of major elements of a broad range of diagnostic activity which will be discussed below. It is important to understanding that the flow chart in Figure 1.1 is a grossly simplified process flow. There are in fact countless branches and decisions in a complete failure analysis process description. In many ways, the process can be viewed as two primary parts: the electrical isolation of the failure and physical cause analysis as shown in the Figure 1.1. The electrical isolation can be viewed as a narrowing of the scope of investigation by determining a more precise electrical cause of failure. For example, a precise electrical cause of failure might be a short between two nets or signal interconnects. The second part of the process is the determination of the physical cause of the failure. This is the process of uncovering what has physically caused the electrical failure cause. For the example above, the physical cause of the failure might be a stainless steel particle, which shorted the two interconnects.

2   Introduction

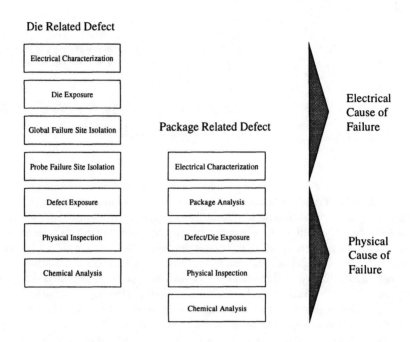

Figure 1.1. Typical failure analysis process flows for die related and package related defects are shown for comparison. Each flow can be broken down in to an electrical cause of failure determination and physical cause of failure determination.

   As pointed out above, the failure analysis process is in fact a part of a larger product or manufacturing improvement process. The input to the failure analysis process is failed devices. In some cases, very specific devices such as qualification failures must be successfully analyzed. In other cases such as yield analysis, where it is not feasible to analyze every failed device, a selection process must occur. The selection is usually made in order to make the biggest possible improvement in the manufacturing process. This typically consists of a Pareto ranking of defect characteristics observed. The goal of the failure analysis process is to enable an improvement process. Hence, the results of the failure analysis, the identification of a physical cause of failure, must be useful for correcting a manufacturing problem. This drives a need for "root cause" identification. This frequently goes beyond the physical cause of the failure. The root cause for the particle, cited in the example above, might be a poorly designed wafer loader for the interconnect deposition tool which results in friction and particle generation.

In simpler terms, the isolation process can be viewed as determination of where to look for the defect or anomaly. The physical analysis process is one of exploration and information gathering about the failure mechanism or physical cause of the failure. The corrective action process is using the physical analysis to understand how the defect or anomaly was generated in the manufacture of the device and how to eliminate the "root cause" of failure.

1.1 ELECTRICAL CHARACTERIZATION

Electrical testing plays a very significant role in failure analysis and is the starting point for every analysis. The initial electrical verification process provides a general understanding of how the device is failing electrically. This is generally in the form of a datalog, the measured continuity, parametric, and functional outputs of the device from production test equipment. This guides the further course of action. The results may drive the need for more detailed electrical characterization. For example, electrical leakage identified may lead to detailed I-V (current-voltage) characterization on a curve tracer or parametric analyzer. For functional failures, observation of functional failures may lead to Schmoo plotting (characterization as function of temperature, power supply voltage or frequency) or scan testing to better understand the characteristics of the failure. This further electrical characterization may provide partial or even complete failure site identification. In highly structured devices such as memories, electrical testing my conclusively identify a failure site as single memory bit which covers less than a square micron. When Design for Test features such as SCAN are present in logic, they can frequently be used to isolate failures a single circuit node or net. The initial electrical characterization can also provide indications of a package-related problem. In such cases, non-destructive analysis of the package may take place before any subsequent activity.

Electrical testing is also required for fail site isolation. While fail site isolation techniques are extremely diverse, they do have two common requirements. They require that the failing device be placed into the failing electrical condition. An understanding of how to place the device into this failing condition is achieved during the electrical characterization. They also require access to the die as discussed below.

1.2 DIE EXPOSURE

Die Exposure is also required to perform fail site isolation techniques since packaging materials are opaque. In addition, direct access to the device is required for failure site isolation techniques. Ceramic packages are normally mechanically opened while plastic encapsulated devices are subjected to a decapsulation process. This is most commonly a jet etch decapsulation. A critical requirement of these processes is to maintain electrical integrity of the device. It is essential that the device continue to fail electrically in the same manner as prior to die exposure. In addition, the external pins and their connections to the die must remain in tact for convenient biasing for failure site isolation.

While in most cases, the active surface of the die is exposed, for some technologies, it is not feasible to expose the active face and maintain electrical functionality. This is observed in most flip-chip devices and in some memory devices where bond wires obscure much of the die surface. An alternative is to expose and polish the back surface of the die and use backside failure site isolation techniques. The second alternative is removing the die completely from package. This results in removing the usual paths for electrical stimulation, which is required for failure site isolation. This can be remedied by repackaging (placing the device in an alternative package better suited to failure analysis and rewiring) or by using a probe card.

1.3 FAIL SITE ISOLATION

Fail Site Isolation consists of a broad range of techniques for narrowing the focus of diagnostic investigation. Modern IC's may have literally many millions of transistors and interconnects. It has become impossible to perform analysis on IC's without narrowing the scope of the failure analysis from the millions of elements down to a very few.

In general, techniques can be classified as global technique or probe techniques. Global techniques attempt to identify secondary effects of the failure. Thermal detection techniques attempt to identify heat generate at a failure site. Photon Emission Microscopy similarly identifies anomalous light emission as a result of electron-hole recombination. A range of other techniques which identify various carrier generated or thermal events are also effective methods of quickly performing failure site isolation.

Probe techniques are those which make measurements of electrical signals within an IC. These techniques can be viewed as direct trouble-shooting of the device by direct measurement of circuit performance. If these techniques are applied without some fail site isolation from electrical testing, the process of fail site isolation by probing becomes a long and tedious one.

1.4 PACKAGE ANALYSIS

Package Analysis is focused on non-destructive techniques for analyzing the structures within an IC package. Electrical data can indicate a possible package related problem, e.g. opens, shorts or leakages. In such cases, it is helpful to assess the physical condition of the package and its interior prior to any destructive analysis. These techniques also initiate the physical cause analysis for open and short failures. Powerful probes such as X-radiography and the Scanning Acoustic Microscopy (SAM) provide particularly useful insights into the integrity of a package structure.

1.5 PHYSICAL AND CHEMICAL ANALYSIS

The task of Physical and Chemical Analysis supports the determination of the physical cause of failure. Once the failure site isolation is complete, analysis must continue to identify the physical cause of failure. There are generally three types of tools employed: sample preparation, physical observation and chemical analysis. Sample preparation techniques are used to remove materials to access the failure site. These include such processes as deprocessing, parallel polishing and cross sectioning. Physical observation is carried using various types of microscopes: optical microscopes, scanning electron microscopes, transmission electron microscopes and others. Chemical analysis is performed on defects where it will provide valuable information in the "root cause" determination.

1.6 DIAGNOSTIC ACTIVITIES

In many ways, the diagnostic process described above has been the standard flow for failure analysis for several decades. The remarkable progress of the semiconductor industry in terms of continuously improved technology at lower cost has been well documented. Many of the industry trends, which have driven this progress, have had a major impact on failure analysis. These include greater device complexity, smaller feature sizes, more levels of interconnect, lower power supply voltages, and package evolution from through hole to surface mount and ultimately to chip-scale packaging. These changes have made the tools required to perform these failure analysis subprocesses much more sophisticated and expensive. These changes have also driven the development of a very diverse set of tools and techniques. This occurs because many of tools work extremely well for one class of defects but are ineffective for others. For example, some global failure site isolation techniques are perform very well for open circuits but provide not benefit on leakage failures. At times, specific tool development has come about because of a specific change in the industry. The use of double level metallization was, in large part, responsible for the development of the global failure site isolation tools such as liquid crystal and photon emission microscopy. Similarly, surface mount technology drove the development of the SAM in order to detect and understand delamination observed during customer assembly of surface mount devices. As the requirements for diagnostic activity increased and became specialized, many of the tools have become specialized to specific applications. For example, in-line tools such as Scanning Electron Microscopes (SEM) and optical microscopes have become highly automated with wafer handling capability and clean room compatibility in order to support wafer fab yield analysis. Similarly, SAM applications have been adapted to allow 100% inspection of devices following assembly in order to assure adhesion of packaging materials to the die.

Just as the toolset has expanded, the application of the failure analysis methodology has also expanded and become customized to an assortment of diagnostic activities. Historically, these tasks were consolidated in the failure analysis laboratories. However as the importance of the various diagnostic activities became understood and the tools became customized, these activities became separate functions.

The subprocesses are used in whole or part in various diagnostic applications: Process Development, Wafer Fab Yield Analysis, Design Debug, Assembly Yield Analysis as well as the Qualification/Customer Return Failure Analysis. The application of the various failure analysis techniques can be looked at a matrix of the technique and the diagnostic activity as shown in Table 1.1.

Diagnostic Activity		Package Analysis	Electrical Characterization	Die Exposure	Failure Site Isolation	Physical/ Chemical Analysis
Process Development Wafer Fab Yield Analysis	Particle Analysis		X			X
	Parametric Analysis		X			X
	Unmodeled Yield Analysis		X		X	X
Design Debug			X		X	
Assembly Yield Analysis		X	X			X
Qualification	Customer Returns	X	X	X	X	X
Customer Return FA	Qualification/ Reliability	X	X	X	X	X

Table 1.1. Diagnostic activities are compared to failure analysis tools used. Electrical characterization is a key element of all diagnostic activities.

These varied applications of failure analysis processes all feed back into to the continuous improvement program of the IC manufacturer.

1.6.1 Process Development and Yield Analysis

In general, these diagnostic activities in the wafer fab can be broken down into three categories. The first category is defect or particle reduction. This is an important element of both process development and ongoing yield improvement. Particle detection tools are used to identify the number, size and location of particles generated at various wafer fab process steps. Physical and chemical analysis is performed both in-line and in wafer fab support areas to determine the physical properties and composition of these particles. However, not all particles generate electrical failures. It is important to be able understand where particle generation is resulting in electrical failures. Much of this process is performed using memory devices or memory imbedded in logic. Memory is highly structured and this allows failure site identification from test data only. In the case of single bit failures, the electrical isolation from test includes identification of the specific memory cell, which fails. Added electrical characterization of the failing bit can identify the specific element of the memory bit that has failed.

The second category of wafer fab diagnostics is parametric analysis. During process development, an understanding of unexpected electrical performance must be characterized and the physical causes understood. This characterization includes both mean value analysis and distribution analysis. During ongoing manufacture, it is essential to track variations in electrical performance and tighten the distribution of key parameters such as drive current, contact resistance, and leakages. These are frequently impacted by process variations, which are not particle related. Where these generic problems exists, test structure analysis should be able to quickly identify the parametric problem. Physical analysis and possibly chemical analysis are used on the test structure.

The final category of diagnostic activity in the wafer fab is the analysis of problems that can not be modeled with the observed defect densities and observed parametric distribution. In the first two categories, the failure site isolation is performed predominant from the electrical characterization either to a memory bit location or to a parametric test structure. This provides a fast and efficient method for providing yield improvement information. The diagnosis of non-modeled failures on the other hand generally utilizes the full range of failure site isolation tools describe above, including both global and probing techniques. Unmodeled yield loss is frequently due to a subtle design to process interaction. This is typically in form of a unique geometric feature.

The diagnostic activities discussed above provide much of the basis for ongoing yield, quality and reliability improvements that have become an expected part of the semiconductor business. The interrelationship between yield, quality and reliability has been intuitively understood for many years. This understanding has recently become more formalized[1]. In addition to the quality and reliability implications, significant profit margin improvement can be achieved through the improvement in yield. For these reasons, much of the increase in diagnostic activity is expected to focus on this area as it has for several years.

1.6.2 Design Debug

Design debug is critical to timely product introductions. When designs do not function as intended when implemented into silicon, it is critical to quickly understand why the device is not functioning properly. It is also critical to assure that all of the defects in the design are corrected in a single pass. Design debug is generally performed through the use of probe measurements on the circuit, circuit analysis and simulations. In general, design debug predominantly uses the probe

8 Introduction

failure site isolation methods from failure analysis although photon emission microscopy can also be useful in the identification of some design defects.

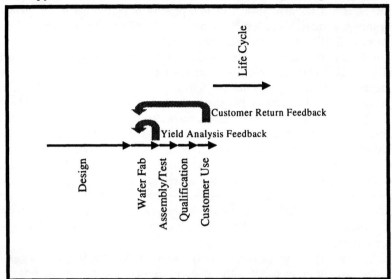

Figure 1.2. Typical feedback loops for customer return and yield analysis on a wafer fab related problem.

A significant trend in the semiconductor industry has been the decrease in the life cycle of IC products. This puts a great deal of pressure on first pass success in the design process. When the first implementation of a design in silicon does not function properly, recovery is critical to the life of the product. Long delays in debug can now easily cause a product to entirely miss its lifetime window.

1.6.3 Assembly Yield Analysis

Assembly yield analysis is a critical part of the ongoing cost reductions in the semiconductor industry. It has an impact on quality and reliability similar to that noted above for wafer fab yield. Assembly yield analysis is generally focused on defects generated in the various assembly processes. Since the purpose of packaging is to provide connectivity between the IC and the external pins, most of the failures associated with the assembly process are opens, shorts and pin-to-pin leakages. Most of these failures are attacked through the use of non-destructive package analysis, predominantly X-ray and SAM. This is inspection process is followed up with physical and chemical analysis.

Just as wafer fab yield is key to semiconductor profitability, packaging yield is also critical. Profits are impacted by the loss of good IC dice. They are also impact by the package costs for any packages, which are assembled but can not be sold.

1.6.4 Qualification and Customer Return FA

The failure analysis of qualification failures encompasses the complete process flow as described above. In many ways, it is the most difficult and diverse of the diagnostic activities since all of the tools must be used at times. In addition, there is a great deal more focus on successful analysis of specific failures. In yield analysis and process development, the goal is typically to understand the statistically significant portions of the yield fall-out. Thus, unique failures are frequently not as critical. In design debug and unmodeled yield loss, the resolution of a specific issue is very critical, but typically a large number of non-functioning devices are available.

Stress	Typical Stress Test	Purpose	Failure Mechanism(s)
High Temperature High Electrical Bias	High Temperature Operating Life Test	Wafer Fab Reliability	Oxide and Metal Wearout Defect Related Failures
Low Temperature High Electrical Bias	Low Temperature Operating Life Test	Hot Carrier	Hot Carrier
High/Low Temperature	Temperature Cycle	Package Reliability	Mechanical Stress
High/Low Temperature Rapid Temp. Change	Thermal Shock	Package Reliability	Mechanical Stress
High Temperature High Humidity/Electrical Bias	85/85	Package Reliability	Corrosion. Delamination
Higher Temperature and Humidity than 85/85	Highly Accelerated Stress Test (HAST)	Package Reliability	Corrosion. Delamination
High Temperature High Humidity	Autoclave	Package Reliability	Corrosion. Delamination
Simulated ESD events	Electrostatic Discharge	Sensitivity to EDS	EOS, ESD

Table 1.2. Typical accelerated stress tests are identified with goals.

Customer return analysis was initially used as product improvement feedback loop. Several factors have slowly eroded that role. The most significant is that this feedback loop is much too long. Figure 1.2 shows a comparison of the feedback loop for customer return failure analysis and yield and analysis. The comparison of the length of the customer return feedback loop to the length of the product lifecycle is also significant. Problems needed be addressed much earlier in the process in order to reduce customer exposure to problems, which occur. The processes, primarily yield analysis, to provide shorter feedback loops have become more effective. Ongoing quality and reliability improvements, Yield Improvement activities, Wafer Level Reliability efforts, and "Maverick" lot reduction activities have provided shorter and more effective feedback loops into the semiconductor manufacturing processes. These activities are expected to reduce the number of serious IC problems to which customers are exposed. Customer return analysis remains critical to support customer business decisions including purges, product recalls, line shut downs when serious problems do occur.

Qualification failures are also critical to the IC business. Qualification consists of a series of accelerated stress tests, which are performed on statistically selected samples from a new device type, new wafer fab process or new package. These test are generally geared towards the detection of know wear-out mechanisms in IC's as shown in Table 1.2. However, they also provide an insight into anticipated use-condition failure rates. When a product or process fails qualification, serious delays in the introduction of the product or process occur. When failures occur, failure analysis is essential to help identify the "root cause" of the failures. Correction of these "root causes" makes qualification on a second pass more certain. In addition, an understanding of the "root causes" may lead to an acceptable screen, e.g. burn-in, for devices already manufactured.

In addition to qualification failures, other failures, which occur on accelerated stress testing, must also be analyzed. While the same stresses as used in qualification are employed, many evaluations and process monitors are performed in addition to qualifications. These provide feedback for ongoing reliability improvements.

1.7 ROOT CAUSE AND CORRECTIVE ACTION

As the diversity and complexity of the failure analysis toolset has increased, the cost has risen as well. With the increased cost, more emphasis is being placed on understanding how a return on investment is generated. The return on investment is through the improvements made during the corrective action process. While this book deals primarily with the tools and techniques of failure analysis, this process can not be considered complete without a "root cause" determination and corrective action. With the challenge of understanding the array of the available tools and techniques for failure analysis, most analysts will have only a general understanding of specific semiconductor manufacturing processes. Since it is not generally possible for the analyst to also have a detailed understanding of the individual semiconductor processes, the task of using the failure analysis results to determine root cause and to generate corrective action lies predominantly with the process engineer or IC designer. The value of the failure analysis is therefore in the use of those results rather than in the generation.

1.8 CONCLUSION

Failure analysis is a difficult task, requiring a diverse set of tools and techniques. In fact, failure analysis itself is a diverse set of applications of this toolset. A critical part of success in failure analysis is tool and technique selection. This is a constantly evolving decision based on several factors. These factors include such things as likely causes of failures based on device history. They also include the probability of success using as particular tool, given a particular electrical signature. Intangible factors such as recent successes or failures with a tool also tend to impact the selection process. This book will provide an understanding of how the various tools and techniques work. Examples will be used to assist in understanding how these tools and techniques can be applied.

REFERENCE

1 Van der Pol JA, Kuper FG, Ooms ER. Proceedings 7th European Symposium for Reliability of Electron Devices, Failure Physics and Analysis (Pergamon Press), 1996, 1603.

2

ELECTRICAL CHARACTERIZATION

Steven Frank
Texas Instruments Incorporated
Wilson Tan
Micron Semiconductor Asia Pte Ltd
John F. West
Cyrix Corporation

This chapter will present the concepts of electrical characterization for the failure analysis of IC's. Electrical characterization plays a key role in the success of failure analysis. It provides the starting point for the complex process of narrowing the scope of analysis from an entire IC to a single bond wire, signal net, transistor or defective component. In some cases, particularly in single bit memory failures, electrical characterization can provide very detailed and precise information about the physical location of the failure. In other cases, a general area of the IC may be indicated. In all cases, electrical characterization provides the failing electrical conditions to perform physical failure site isolation described in later chapters. While most failure analysis tools and techniques bridge the full range of IC's, significant differences exist in methodology and utilization of electrical characterization tools for different types of IC's. In general, we can view the differences in the three broad categories of IC's: logic, analog and memory. These distinctions may appear blurred as we approach "system on a chip" devices which contain all of these elements. Even when merged, however, there are elements of electrical characterization for these three types of circuits that carry through to the component blocks.

In today's world, logic IC's encompass a broad range of semiconductor devices. The most common are microprocessors, microcontrollers, and digital signal processors. Other logic devices range from "glue" logic chips, which may consist of only a few gates to Application Specific IC's (ASIC's) which may contain hundreds of thousands of gates. ASIC's have become very complex and may contain embedded blocks of memory and analog functions in addition to digital logic. Determining the electrical cause of failure in one of these devices is not a trivial task and electrical characterization provides the first valuable clues in the failure site isolation process.

Memory, whether embedded in a logic device or stand alone, provides the most highly structured of all IC's. These highly structured circuits are very observable and

14 Electrical Characterization

can typically be characterized very precisely. In logic devices with embedded memory, the memory blocks are commonly used in wafer fab yield analysis to drive defect detection because of the ease of failure site isolation. However, logic content in memory devices is rapidly growing to support its interaction with processors. This results in a need to use more "logic-like" failure analysis techniques for memories.

Even as the world rapidly becomes digital, interface to analog inputs and outputs remains a significant requirement. The increase in mixed signal (analog and digital) integrated circuits reflects this. Characterization of analog IC's and analog circuit blocks is particularly challenging because failures tend to occur with small deviations in voltage and current.

The topics discussed in this chapter include basic equipment and techniques used for circuit characterization, electrical characterization for failure site isolation, and future challenges for circuit failure isolation.

2.1 ELECTRICAL CHARACTERIZATION

Electrical failures, in general, can be divided into three areas: continuity, parametric and functional failures. Each of these types of electrical failures has its own requirements for characterization. Continuity failures are the easiest to characterize and only require a simple measurement for opens and shorts on the external device pins. A short may also be further characterized as to its measured resistance.

Parametric failures require more complex measurements for characterization. These might include input or output voltage levels, power supply current, bias currents, offset voltages, overcurrent shutdown levels, frequency response, or other parameters specific for a given device. Part of the complexity with characterization of parametric fails is that each circuit will have its own distinct parameters, with each parameter specified to fall within a range of acceptable values. While most parameters for devices tend to be common across the product families, each family of devices has parameters unique to that family. For example, op-amps will have different specified parameters than an analog to digital converter might have. Parametric characterization is also used to measure individual component parameters such as bipolar transistor DC current gain, MOS transistor threshold voltage, capacitor matching, or resistor matching.

Functional failure characterization is performed by inputting a known stimulus and measuring the resulting outputs. If the measured output states correspond to the expected output states, the part is said to be functional. If not, the part is failing functionally. Functional testing over a range of supply voltages or temperatures may give clues to help isolate failures (known as shmoo plotting). At times, the distinction between parametric and functional fails can blur. For example, power supply current parameters, commonly referred as I_{DDQ} in CMOS devices, is a unique parameter in that it may require creating a very specific functional state in order to measure the leakage current. If the failure is functional state independent or easily achieved, I_{DDQ} failures are more like parametric failures. When the electrical stimulus becomes very complex, their characterization is more like that of functional failures.

2.1.1 Tools for Electrical Characterization

The most basic tools for electrical characterization are those used for continuity testing and basic parametric analysis. For these tasks, a curve tracer is an essential tool. A curve tracer applies a variable voltage (V) to the device under test (DUT) and displays the resulting current flow (I) in an X-Y plot. It is used to examine the I-V characteristics of device I/O (input/output) pins and can also be used to perform parametric characterization of internal resistors, diodes, and transistors. For example, the curve tracer[1] can be used to measure the DC current gain (β) of bipolar transistors as well as act as a load to measure output parameters such as V_{OL}, V_{OH}, and V_{SAT}. The use of specially constructed switching boxes for low to medium pin count devices facilitates measurement from various pins. A manual curve tracer has the advantages of high available output current, versatility, and ease of use. It does, however, have disadvantages that severely limit its usefulness for I/O pin characterization of high pin count devices. A manual curve tracer has no facility to fixture high pin count devices. It also lacks the ability to store I-V curves for a good-bad comparison of I-V curves.

A computer controlled curve tracer overcomes the limitations of the manual curve tracer for high pin count devices and has become a standard tool for I/O electrical characterization. The computer controlled curve tracer standardizes fixturing with a standard zero insertion force (ZIF) socket (e.g. 21 X 21 PGA) that is connected to analog bus lines via an electrically addressable switching matrix. The computer controls the application of the stimulus to each pin and records the measured data for display or comparison to previously stored data. The analog busses are configurable so that it can be setup for standard curve trace analysis, powered curve trace analysis, or latch-up testing [2,3].

The tools required for electrical characterization of parametric failures can vary greatly depending on the type of parameter that failed. A semiconductor parameter analyzer is a useful tool for characterizing diodes and transistors, performing circuit level parametric testing which includes DC functional testing of analog circuits such as voltage references, regulators, and op-amps. A semiconductor parameter analyzer is an instrument that is made up of programmable I-V source/monitor units that can be independently configured in a variety of ways. Semiconductor parameter analyzers can measure very low levels of current (fA resolution) and voltages (µV resolution), making them very useful for a broad range of measurement and characterization functions.

Logic devices parametric failures fall into two categories, logic-state independent and logic-state dependent. Logic state independent failures are those which can be directly observed from the external device pins regardless of the state of the other pins. In some cases, such as V_{OH}, the characterization of the failure requires placing the output into a high state through the logic of the internal circuitry. The

16   Electrical Characterization

requirements to achieve a high state may vary from a simple DC level of a set of input pins to a complex set of functional test vectors. The simpler device failures are normally analyzed in a lab using bench-test equipment such as a switch box, stimulated with power supplies, pattern generators, curve tracers, and parameter analyzers. Complex devices normally cannot be characterized on a bench setup due to the large number of pins, which must be controlled simultaneously. They require a higher level of test equipment for the electrical stimulus. In this regard, state dependent parametric failures can become more like functional failures in terms of the electrical stimulus required to create them.

The tools required to characterize functional failures will also depend on the type of failure encountered. Functional failures on simple devices can be analyzed using bench test electronics such as power supplies, function generators, pattern generators, and logic analyzers. As devices become more complex, the number of vectors required to achieve the desired state (vector depth) becomes much higher and the timing of the inputs becomes more critical. For many years, failure analysis performed this function with a class of testers, frequently categorized as ASIC verification testers. These testers filled a gap between bench pattern generators/logic analyzers and production test equipment. These testers, compared to production versions, were of lower cost. Further, conversion of test patterns for low pin count devices was relatively easy but did not assure correlation to production ATE (Automatic Test Equipment). As the number of I/O pins and timing complexity on devices have increased steadily, this correlation with production ATE has become a more significant issue. As pin counts in all areas continue to increase, the use of ATE as a tool for electrical characterization is expected to increase. Factors such as increased number of power supplies, increased timing complexity, analog content and vector depth make the ASIC verification tester less suitable for failure analysis purposes.

The ATE also provides a software toolkit for debug and failure analysis through its graphical user interface. The most common debug tools available are wave, vector or pattern, and shmoo tools. The wave (or scope) tool is a digitizing sampling oscilloscope that displays both input and output waveforms of the device. For outputs, the true waveform can be displayed along with trip levels and output strobe markers. The vector or pattern tool displays the inputs and outputs for the entire vector set. All failing vectors and the failing pins within the vector are highlighted. The most useful characterization tool is the shmoo tool. A one, two, or three dimensional shmoo plots allow multiple parameters to be simultaneously varied in order to determine the conditions under which the device passes and fails. While using these tools, device timing and logic level definitions can be modified "on the fly", eliminating the need to recompile vectors and the test program.

The use of ATE for failure analysis provides several distinct advantages. Since the tester is used in production, verified test programs already exist. No correlation issues should occur. In addition to having existing test programs, test hardware is also readily available. Typically, a failure analysis compatible test socket is the only added hardware cost. The tester has been matched to the product to assure that it is capable of adequately testing the device. The tester cost of ownership is also significantly reduced. Since the tester is not dedicated to failure analysis, it can be utilized for production or engineering when not employed for failure analysis. Cycle time is also positively impacted. Since the hardware and test programs are debugged for production, little or no effort is expended in correlation and debug activitiesIn addition, the production testers are typically well maintained and calibrated for production activity.

2.1.2 Electrical Characterization for Fault Isolation

Failure analysis normally begins with an examination of the datalog submitted with the failing unit. A datalog is the measured results of the production test program. The datalog submitted with a failing unit should be a full datalog, i.e. a listing of the results of all of the production tests. Most production test programs stop at the first indication of failure and this feature must be overridden to generate a full datalog. It may also include data for device operation at various temperatures and operating voltages. Interpretation of this data is a key factor in efficient failure analysis.

By examining all the fails, the analyst can use a heuristic approach to determine if one failure mode (an observed electrical failure) is dominant and causes the others to occur. It is common for a failure to have a large of number of failures documented in a datalog. For example, a device fails functional testing across the V_{DD} range and an input leakage high (I_{IH}) test on one input pin with mA of leakage (a normal reading is on the order of nanoamperes). In this case, the leakage is likely the cause of the functional fail because it is preventing the input buffer from reaching the desired V_{IH} with the appropriate timing. This prevents the pin from switching properly. Analyzing the failure as a functional fail would add significantly to the complexity of the analysis. In general, the failure mode should be expressed in the simplest terms for failure analysis. If many different tests are failing the test program, the sequence of failure mode types in order of ease for failure analysis is: continuity, parametric, functional in that order.

Continuity failures typically cause a large number of failure modes. A device failing with an open input will drive many additional failures, functional as well as I_{DDQ}, due to improper conditioning and floating gates, etc. Opens on outputs will cause functional and V_{OUT} parametric fails. Shorts creates both functional and I_{DDQ} failures, as well as leakage and V_{OUT} parametric tests. Power supply shorts are typically seen as I_{DDQ} and functional fails.

Parametric failures can also result in functional failure modes. For example input leakage may cause I_{DDQ} to fail as well as slowing the input enough to cause at-speed functional failures. Conversely, functional failures can lead to parametric failures. Many output parameters are dependent on creating a particular state on that output. Thus, a true functional fail may fail V_{OUT} and I_{OZ} tests in addition to the functional patterns. The difference can usually be established by determining the state of the

18   Electrical Characterization

output at the time the parametric test is performed. Thus, if the output is in correct state, V_{OUT} and I_{OZ} failures can be real.

In general, a thorough evaluation of the datalog will allow the reduction of the analysis to that of a continuity failure or a simple parametric analysis where that is possible. In addition to ascertaining what test is failing, it is important to understand the magnitude of the failure. This drives subsequent failure site isolation. For example, the approach on a 1 milliampere I_{DDQ} failure is likely to differ from a 1 microampere I_{DDQ} failure. If available, a datalog should also be acquired from a passing or correlation device. This will give the analyst a good indication as to what to expect for typical parametric values.

2.2 CURVE TRACING

Curve tracing is an important technique that is used to measure the electrical characteristics of the I/O structures on integrated circuits. A curve tracer or parametric analyzer is used to apply a variable voltage to a device pin and display the resulting current flow as a function of the voltage. This I-V curve is a characteristic curve, and any deviations from an ideal I-V curve will give clues for possible failure mechanisms. Curve tracing is also useful in characterizing internal device components, such as resistors, diodes, and transistors. This section presents the use of curve tracers and parametric analyzers in the characterization of diodes, transistors, and I/O structures.

2.2.1 Diode Characterization

The most basic building block for an integrated circuit is the PN junction. The PN junction forms the basis of diodes and transistors and an understanding of its characterization and possible failure mechanisms is crucial to understanding structures that are more complex. A common measurement for a PN junction diode is the current through the diode as a function of the applied voltage (the positive terminal is taken to be the P side). Figure 2.1 shows the I-V curve for a PN junction diode and identifies its three regions of operation. The forward bias region occurs when the applied voltage is greater than zero. In this region, the current flow through the diode is an exponential function of the applied voltage. The second region of operation is the

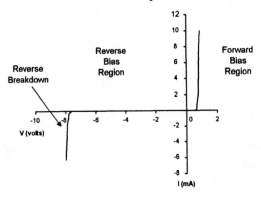

Figure 2.1. I-V characteristics of a PN junction showing the three regions of operation.

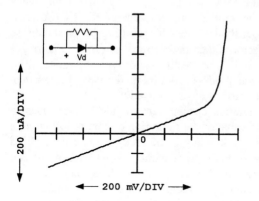

Figure 2.2. An example of a failed PN junction that is electrically equivalent to a resistor in parallel with the junction is shown.

reverse bias region. Here, the applied voltage is less than zero and the current through the diode is a small constant value (essentially zero) and is called the reverse saturation current. The third region of operation is the reverse bias breakdown region. In this region, the breakdown voltage of the PN junction is reached and the current through the junction increases rapidly due to avalanche multiplication. This current needs to be limited by an external resistor or permanent damage to the junction will result.

Measured deviations from the normal I-V curve of a diode can provide an indication of how the diode may have failed. For example, a low value resistance shunt in the I-V measurement can indicate a catastrophic and irreversible degradation of the junction. This type of degradation is commonly caused by an electrical overstress (EOS) condition. Figure 2.2 shows an I-V curve of a damaged PN junction whose electrical properties were equivalent to a resistor in parallel with the junction. If the damage to the junction were more severe, a short would have resulted.

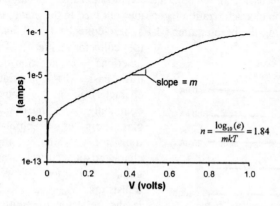

Figure 2.3. $Log_{10}(I)$ vs. V for a discrete diode. The ideality factor is calculated from the slope of the line.

Characterization of a PN junction in its three regions of operation can provide helpful clues to possible failure mechanisms. In the forward bias region, a parameter called the diode ideality factor, n, is one that can be used to detect junctions that have been electrically overstressed. It can be extracted from a modified I-V where the current through the diode is plotted on a logarithmic scale. This can easily be done using a parameter analyzer. Recall that in the forward bias region the current through a diode is an exponential function of the applied voltage. Therefore, if the current through the diode is plotted on a logarithmic scale, the resulting curve should be linear with a slope m. The ideality factor is then proportional to $(1/m)$. Figure 2.3 shows how the ideality factor is measured. The ideality factor typically ranges from 1 (for integrated diodes) to 2 (for discrete diodes) and a measured value greater than 2 indicates that the junction has been electrically over-stressed[4,5].

In the reverse bias region, the saturation current can provide helpful clues to possible failure mechanisms. Since the value of the saturation current is so low, it is very sensitive to damage that result in low levels of leakage current, such as electrostatic discharge (ESD) damage. This type of damage is easily isolated and identified. The reverse saturation current is also sensitive to surface conduction and therefore a high reverse saturation current may be indicative of ionic contamination.

Finally, the reverse breakdown voltage can provide helpful clues about possible failure mechanisms. First, the value of the breakdown voltage is dependent on process parameters and a low breakdown voltage may point to a process anomaly. However, it also may be due to an electrical overstress condition that has damaged the junction. An erratic and unstable reverse breakdown I-V curve may indicate a cracked die. Finally, characterizing the reverse bias breakdown voltage can uncover a phenomenon called 'walkout'. Walkout is a term that is used to describe the upward drift of the reverse bias breakdown voltage with increasing applied current. Walkout results from surface avalanche and is due to charge injection into the oxide above the silicon surface[1,6].

2.2.2 Transistor Characterization

This section discusses the characterization of bipolar and MOS transistors with emphasis on parameters that are useful in detecting failure mechanisms. In bipolar transistors, the basic PN junction characterization techniques outlined in the previous section are used in characterizing the base-emitter (B-E), base-collector (B-C), and the collector-emitter (C-E) junctions. In addition to this basic PN junction characterization, some commonly measured parameters of a bipolar transistor are: the characteristic curves, the gain, and for power transistors, $V_{CE(SAT)}$, which is the collector to emitter voltage when the transistor is in saturation. The characteristic curves for a

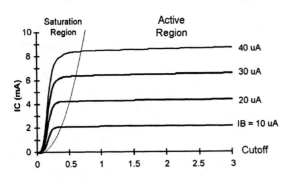

Figure 2.4. Characteristic curves for an NPN transistor showing the three regions of operation.

bipolar transistor are a family of plots of the collector current (I_C) versus the collector-emitter voltage (V_{CE}) for a series of base currents (I_B). An example for a NPN transistor can be found in Figure 2.4. Transistor parameters such as gain (β), the Early voltage (VA), and the collector-emitter breakdown voltage (BV_{CEO}) can be determined from these characteristic curves. Each of these parameters can provide helpful clues to possible failure mechanisms.

The Early voltage is inversely related to the variation of I_C with V_{CE}, which is merely the slope of the characteristic curve in the active region. A low value for the Early voltage means that the collector current of that particular transistor is dependent on the value of the collector voltage. This is electrically equivalent to having a resistor from collector to emitter.

The collector to emitter breakdown voltage (BV_{CEO}) is analogous to the PN junction reverse breakdown voltage discussed earlier. The failure mechanisms found in the PN junction analysis section also apply here.

The DC current gain (β) is the second major parameter measured in conjunction with failure analysis of the bipolar transistor. β is defined as the ratio of the collector current to the base current and is measured in a variety of ways. In the proceeding section, it was stated that the gain of a transistor could be found by its characteristic curves. It also can be measured directly on a parameter analyzer by configuring it to sweep the base current while measuring the collector current. The measurement of β for a transistor can provide helpful clues about possible failure mechanisms. Degradation of β at low collector currents is an indication that the transistor has been electrically stressed, typically by biasing the base-emitter junction into avalanche. The degradation is attributed to surface recombination within the E-B depletion region at the oxide interface[7,8,9]. Other characteristics of this failure mechanism that can be measured are: the transistor

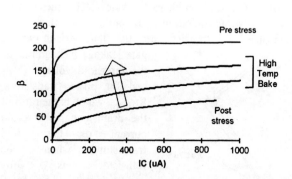

Figure 2.5. Variation of β with IC before and after reverse bias B-E avalanche is illustrated. Pre-stress, post-stress, and bake-recovered (125°C) gain behavior is shown.

will have normal β at high collector currents, the transistor will have normal B-E diode characteristics, and the gain will recover after a high temperature bake. In analog circuits, β degradation adversely affects input bias currents and input offset voltages. Figure 2.5 graphically shows the affect of reverse bias B-E avalanche on the β of a transistor. It shows clearly the degradation of β after stress, and it's partial recovery after bake.

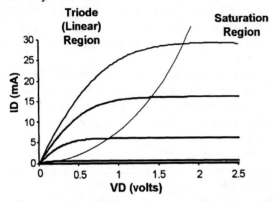

Figure 2.6. Characteristic curves for a NMOS transistor show the regions of operation.

The collector-emitter saturation voltage ($V_{CE(SAT)}$) is the third bipolar transistor parameter commonly used in failure analysis, particularly for high current output structures. In output structures, $V_{CE(SAT)}$ is the voltage measured at the output pin when the output power transistor is fully conducting. The parameter is particularly sensitive to series resistance values introduced by assembly; namely ball bond and stitch bond resistance.

CMOS processes have become dominant in the industry, in not only digital circuits, but analog circuits as well. The following sections will discuss the characterization of MOS transistors, with emphasis on parameters that are useful in detecting failure mechanisms. Basic PN junction characterization techniques can be used for the drain, source, and substrate junctions. In addition to this, some commonly measured parameters of a MOS transistor are: the characteristic curves, gate leakage measurements, the threshold voltage (V_T), and for power transistors, $R_{DS(ON)}$, which is the ratio of the drain voltage to drain current of the MOS transistor when the transistor is fully conducting. The MOS characteristic curve is similar to the bipolar curve and is an important source of information. The characteristic curves for an NMOS transistor is shown in Figure 2.6 and are a family of plots of the drain current (I_D) versus the drain

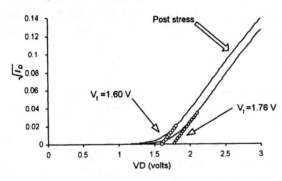

Figure 2.7. The V_T shift is shown for a NMOS transistor before and after a gate stress illustrating variation in V_T due to charge injection into the gate oxide (65 μA of current gate current for 2 seconds with the drain and source grounded).

voltage (V_D) for a series of fixed gate voltage (V_{GS}). The possible failure mechanisms detected with the MOS characteristic curve are similar to those found with the bipolar curves. A MOS transistor has a parameter that is analogous to the Early voltage and a low value on this parameter is electrically equivalent to a resistor from drain to source.

The second common parameter measured on MOS transistors is gate leakage. Gate leakage in a MOS in transistor is current flow between the gate and either the source or drain. Normally, there is no current flow into the gate and leakage can result in device failures. Leakage can result in I/O parametric leakage failures, power supply current failures, input bias current failures and in some cases, functional failures. Gate leakage measurements are a simple procedure using a curve tracer or semiconductor parameter analyzer. A standard I-V measurement set-up is all that is required to measure and characterize gate leakage. Failure mechanisms responsible for gate leakage include gate oxide defects, electrical overstress damage to the oxide, and certain process defects.

The third common parameter measured on MOS transistors is the transistor threshold voltage. The threshold voltage for a MOS transistor is simply the gate voltage corresponding to the onset of strong inversion in the channel. There are several different ways to measure the threshold voltage, but a commonly used method in failure analysis is to plot to square root of the drain current versus the gate to source voltage[10]. With this kind of plot, the curve will be a straight line and the threshold voltage will be the X-intercept. Figure 2.7 shows how the threshold voltage is found using this method. Figure 2.7 also shows one of the failure mechanisms associated with the threshold voltage shifts. Shifts in threshold voltages will have a detrimental effect on circuits that require a high degree of transistor matching for proper operation. Shifts in threshold voltages occur in a variety of situations but the basic mechanism is the same - unwanted charge in or near the gate oxide[11]. Two of the common reasons for threshold voltage shifts are ionic contamination and charge injection into the gate oxide caused by an overstress condition or hot channel carriers.

Finally, R_{DSON} is a parameter commonly measured for MOS power transistors. RDS_{ON} is analogous to $V_{CE(SAT)}$ in bipolar transistors and is equal to the ratio of the drain voltage to the drain current. Like $V_{CE(SAT)}$ for bipolar transistors, it is highly sensitive to series resistance and assembly related issues.

2.2.3 I/O characterization

I/O characterization is primarily performed by curve tracing each pin of an unpowered device or performing powered curve tracing to measure input current parameters I_{IH} and I_{IL}. I_{IH} is the current into the pin when a logical 'high' voltage is present. IIL is the current out of the pin when a logic 'low' voltage is present. Figure 2.8 shows a simplified schematic of a typical I/O pin. The ESD protection circuitry presents the electrical equivalent of two diodes at the external pin. One diode is reverse biased to V_{DD} and one diode is reverse biased to ground under

24 Electrical Characterization

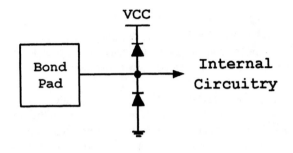

Figure 2.8. Typical I/O pin equivalent circuit is shown.

normal operating conditions. This makes I/O characterization similar to diode characterization.

One standard approach to I/O characterization is to curve trace each pin with respect to ground, power supplies, and adjacent pins. In automated curve tracers, the plots for each pin to ground or a power supply can be acquired and overlaid. Since all signal pins have essentially the same expected characteristic, any deviation is readily recognized. Continuity failures and most I/O related failures lead to an anomaly in the I-V curves on these pins.

Curve tracing can be used to characterize output parametrics if the device can be powered into the desired state, tri-state (Z) or valid output (L or H) condition. For example, tri-state leakage (I_{OZ}) testing with a curve tracer is simply a curve trace of the I/O while the pin is in the Z state (commonly referred as powered curve tracing). The resultant curve trace should indicate infinite impedance between the regions in which the ESD diodes are forward biased.

If a basic curve trace does not indicate leakage on pins, which fail output parameters such as V_{OH} (voltage output high) and V_{OL} (voltage output low), a different approach is used to detect abnormal series resistance on the failing pin. The device is powered to place the failing output pin in the correct logic state (L or H). The failing pin is loaded with the appropriate current source value (e.g. 2mA, 8mA) and the out voltage is measured using a curve tracer or bench test setup. For example, a 5 volt device with an 8 mA output must exhibit a V_{OHL} (voltage output high loaded) greater than 3.7 volts. By sourcing 8 mA to ground on the failing pin, an abnormal series resistance is indicated if the voltage measured at the failing pin is below 3.7 volts. In a similar manner, V_{OL} failures are characterized by sinking the rated current into the failing pin and observing a voltage above the minimum specification limit.

An alternate procedure for curve tracing pins is to ground every pin on the device and then remove one pin at a time to curve trace it with respect to ground, typically with a voltage range of -2 to 2 volts. The current is clamped at a level so as not to aggravate any damage that might be present. The measured I-V curve using this approach will be two forward biased diodes, one in the negative direction corresponding to the diode between the I/O pin and ground, and one in the positive direction corresponding to the diode between the I/O pin and V_{CC}. This procedure has the advantage of reducing the number of curves needed for ground and power supply measurements

2.3 ELECTRICAL CHARACTERIZATION OF STATE DEPENDENT LOGIC FAILURES

As stated in the introduction, logic failures can be divided into two categories, state dependent and state independent failures. State independent failures are typically analyzed as continuity or parametric failures as described in a previous section of this chapter. Since no specific logic state is required to create the failing condition, electrical stimulation for failure analysis is relatively easy to setup. State dependent failures are those in which the device needs to be conditioned into a particular logic state to generate the failure. In general, these include most I_{DDQ} failures, functional failures, and certain output parametric measurements (Many output parametric failures require creating a particular state condition on the output. Since the input conditions to achieve this logic state may be quite complex, they must be treated as state dependent failures from an electrical stimulus perspective but behave more like parametric failures during failure site isolation.).

While most logic state independent failures are readily observable from a powered or unpowered I-V characterization of the external pins of the device, logic-state dependent failures typically manifest themselves within the core of complex logic devices and are much more difficult to stimulate. The failure modes which predominantly fall into this category are I_{DDQ} and functional. Since these types of failures are the most challenging to characterize and isolate, it is important to take advantage of test information and testability features of the device such as I_{DDQ} and scan. Without testability features, I_{DDQ} and scan being the most common, the best case is that the vectors which are failing indicate a block circuit in which to focus failure site isolation activities, making failure site isolation much more difficult. I_{DDQ} is a particularly powerful tool for failure analysis as well as device testing of CMOS logic with low power consumption. Scan provides a method for improving both the controllability and observability of internal circuit nodes. Both controllability and observability are important factors for failure site isolation.

2.3.1 I_{DDQ} Testing

I_{DDQ} is a particularly important parameter for failure analysis. This leakage current, which is frequently present on devices categorized as functional failures, can lead to failures being isolated by global techniques rather than probing techniques. I_{DDQ} is defined as the current flowing from the power supply to ground when the device is in a quiescent state[12]. In a quiescent or static state, N and P channel transistors are either "on" or "off", and in the absence of through current devices or defects, there is nominally no active current. Transistors, which are "on", are in saturation and driving either a logic "0" or a logic "1". In the I_{DDQ} test methodology, vectors are executed to a carefully chosen locations and halted at a parametric measurement stopping place (PM stop) to measure various parameters such as the output parameters and I_{DDQ}. PM stops are normally selected to provide low I_{DDQ}. This is in fact the case for most CMOS static logic. High I_{DDQ} can result from circuits, such as dynamic logic and memory, which do not operate at zero nominal standby current. Sub-threshold (I_{OFF}) transistor leakage[13] is the primary contributor to the background leakage current. Outside of the very deep submicron regime, this leakage is extremely low compared the leakage level required to cause a circuit failure. As sub-threshold leakage increases in the deep submicron regime, I_{DDQ} testability will become limited due to the high level of background current. I_{DDQ} is particularly strongly impacted by the sub-threshold leakage because normal I_{DDQ} represents the sum of the sub-threshold leakage for every transistor on an IC.

From a production test program perspective, many I_{DDQ} PM stop measurements are required to detect all possible faults. With each additional PM stop location within the vectors, a higher percentage of nodes within the device are toggled and checked for a leakage path from V_{DD} to Ground. On the other hand, PM stops consume a significant amount of test time, limiting the number of stops used in production test programs. For failure analysis applications, the test time limitation is not as relevant and I_{DDQ} can be measured at more points. For example, I_{DDQ} measures could be made at a failing vector and vectors immediately preceding the failing vector. In addition, I_{DDQ} can be characterized in more detail as a function of V_{DD}, time and temperature for a failure analysis test program.

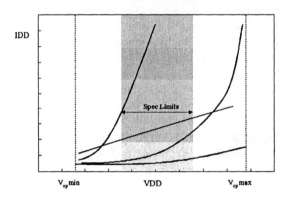

Figure 2.9. The shape of the curve for I_{DD} vs V_{DD} can indicate likely failure mechanisms.

Logic devices can have many power supplies. It is important to understand which power supply exhibits I_{DDQ} leakage. This can eliminate large parts of the circuit as possible failure sites since different power supplies are commonly used in different sections of the logic device. Correlation units are an important part of I_{DDQ} failure analysis. For devices, which do not exhibit low background I_{DDQ}, the correlation units are used to identify the location of the stand-by current so that it is not incorrectly identified as a failure site. Correlation units are also important in the understanding of I_{DDQ} as a function of V_{DD}, time and temperature. If some of the PM stops are passing, they can often be used for correlation purposes.

2.3.2 Voltage Dependency of I_{DDQ}

The most valuable characterization method for I_{DDQ} leakage failures is the generation of plots of I_{DD} versus V_{DD}. In most CMOS technologies, the defect leakage current is typically several orders of magnitude greater than the background leakage (i.e. assuming negligible transistor sub-threshold leakage). A good device should have little or no I_{DD} current until V_{DD} approaches the BVDSS (reverse breakdown voltage) of the CMOS transistors. With background low, the I_{DD} versus V_{DD} plot is essentially a powered curve trace of the defective area of the IC. This can provide significant information about the nature of the defect (see figure 2.9).

A linear characteristic suggests an ohmic defect such as a metal bridge caused by particles or incomplete metal etching. A saturation I_{DD} versus V_{DD} characteristic suggests that a transistor, which is supposed to be "off", is actually "on" and is sourcing current between V_{DD} and ground. This forms a broad class of failure mechanisms. A shallow slope may indicate diffusion spacing violations, oxide microcracking, or filaments. Other curves vary as the square of the V_{DD} voltage indicating a transistor current. These curves are used to determine which failure isolation technique needs to be used in the resolution of the problem.

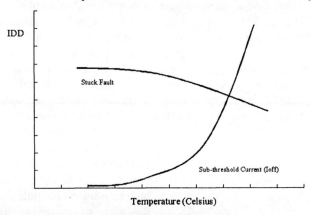

Figure 2.10. The temperature dependence of I_{DDQ} can indicate a stuck-at or leakage mechanism.

Since the I_{DDQ} current is state dependent, the I_{DDQ} characterization must be taken by adjusting voltages on the device, which must remain in the defined logic state. If V_{DD} is reduced below the minimum operating voltage of the functional patterns, the device may lose its state. Similarly, if the input voltages are varied below the V_T of the input buffer, the logic level of some of the internal nodes may change to a different state effectively creating a different I_{DDQ} PM stop. Fortunately, the tools available on the ATE make this characterization a fairly easy process. The measurement can be recorded manually by using software tools, or they can be programmed into the test program so the I_{DD} versus V_{DD} results can be automatically printed or plotted. The execution of the test program is halted at the failing I_{DDQ} PM stop, maintaining the power supply settings and input drive stimulus to the device. While halted at the failing PM stop, V_{DD} voltage is varied across the operating range of the device (the minimum and maximum operating V_{DD} of the device are measured using the functional debug tools on the failing device or a correlation device) and the I_{DDQ} value is recorded as function of V_{DD}. Note that the input drivers must vary with V_{DD} to prevent forward biasing V_{DD} diodes used in ESD protection structures.

28 Electrical Characterization

2.3.3 Temperature Dependence of I_{DDQ}

Recording the response of the power supply current across the minimum and maximum temperature range of the device also provides valuable information about an I_{DDQ} failure. I_{DDQ} for each failing PM stop is measured over a range of temperatures, using standard ATE temperature tools. I_{DDQ} for a stuck fault defect is expected to decrease current with increasing temperature due to the reduction in transistor saturation current at elevated temperatures. I_{DDQ} due to junction leakages is expected to increase rapidly as the temperature is increased. Sub-threshold current (I_{off}) on narrow gate devices typically doubles for every 10 degrees Celsius increase. These typical temperature behaviors are illustrated in Figure 2.10. Thus the temperature dependence of I_{DDQ} can provide insight into likely failure mechanisms.

2.3.4 Time dependent I_{DDQ}

The test program usually makes an I_{DDQ} measurement within 50ms after halting the vectors at the specified PM stop. The measurement is delayed in order to allow a stable measurement of the anticipated low current. Most defects have no time dependency and the current should stabilize at a given V_{DD} very quickly. However, floating gates generate a time-dependent I_{DDQ}. If a gate initially floats to a voltage of one-half of V_{DD}, both n-channel and p-channel transistors in standard logic will be partially on and conduct current from V_{DD} to ground. Eventually, the floating node will charge to a logic state and the I_{DDQ} current will subside. This behavior can be generated by a design defect or a via, contact or metallization line which is very resistive or open. For high resistances, the time for the node to charge to its proper level will typically be long enough for the production test program to record an elevated I_{DDQ} current. In Figure 2.11, a plot of I_{DDQ} versus time for a device having a resistive via is shown.

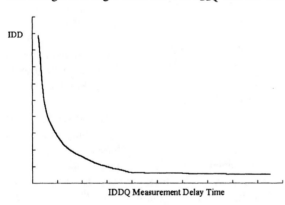

Figure 2.11. A typical time dependent IDDQ is shown.

The data was collected on an ATE by inserting a looping procedure into the program and adding a specified wait time before the next measurement was made. If a device displays this I_{DDQ} instability, the I_{DDQ} PM stop vectors can be executed in a tight loop during failure isolation in conjunction with a photon emission microscope to isolate the failure.

2.3.5 Functional Pattern Failures and Structured Design for Test (DFT)

Traditionally, a failure in the core of the device has been detected with functional vectors. These vectors can be created with an ATPG (Automatic Test Program Generation) synthesis tool or modeled on the circuit application. Measurement of the quality of the functional patterns is based the percentage of nets within the design for which toggling between a logic "0" and a logic "1" (based on a stuck-at-fault model) can be detected. Functional patterns serve various functions. "Loose functional" patterns are typically run at low frequency and are intended to verify gross functionality. "At-speed" vectors typically test the chip's application vectors at the maximum frequency intended for operation. "Delay fault" vectors guarantee critical timing to components external to the IC making a measurement between various input and output timing edges. These vectors are intended to verify critical speed paths in the circuit.

In some cases, specific functional vectors are tailored to test a particular block of circuitry. Thus specific failing vectors can potentially isolate the failure to a particular area of the IC. . In the worst case, functional tests define only the failing output from which to backtrace the failure.

As designs become more complex, it becomes much more difficult to verify with electrical test that logic IC's are completely functional. DFT methodologies are typically required to break the IC into more manageable blocks of circuitry. In order to test devices in this way, the inputs to the reduced circuits must be controllable and the output of the blocks observable. Scan is the most common method used to improve the observability and controllability of a logic design. Using scan insertion software tools, existing flip-flops are modified to have a "test mode" of operation, permitting serial access to the flip-flops. ATPG software tools generate vector sets with a high fault grade. Scan vectors allow the internal logic to quickly obtain a known state. In this manner, a device can be tested with much fewer vectors than conventional functional vectors[14].

In addition to improving test coverage, scan can be successfully used to isolate failures, using the circuit controllability and observability[15,16]. A single or combination of passing and failing scan vectors can be correlated back to specific nodes within the design. Characterization of these failures often requires running a complete set of scan vectors, rather than exiting the vector set on the first fail as in the production test program. Some ATPG tools include software diagnosis. The diagnostic software predicts nodes or nets within the die that may have a stuck fault. In ideal situations, faults can be isolated to a single node or set of equivalent nodes. In any case, it significantly reduces the area of the die to be considered in failure site isolation.

In order to run diagnosis and to predict failing nets within a scan design, the scan chain must obviously be intact and operational. Nearly all IC designs will implement a scan flush or scan check vector to verify scan chain operation. Defects can occur in the scan chains as well as in the functional logic circuit. Since the scan chain is

30 Electrical Characterization

serially connected flip-flops, the chain can be checked with a relatively simple binary search to find the broken chain using probes.

2.3.6 Functional Failure Characterization

A hard functional failure is defined as a device, which fails consistently over different voltages, temperatures, frequencies and input timing ranges. These typically are the result of defects that create true stuck-at faults within the device. A soft functional failure may change with voltage, temperature, frequency or input timing. The device may fail a vector set at one V_{DD} or temperature, but may either pass or change cycles (vector depth) and pins at a different voltage or temperature. Soft failures are often due to an internal circuit node, which is not switching with the correct timing. This can occur due to a delay introduced into a signal path or leakage, which slows a rise or fall time. Characterization of the dependency of soft failures can provide useful insight into the likely failure mechanisms.

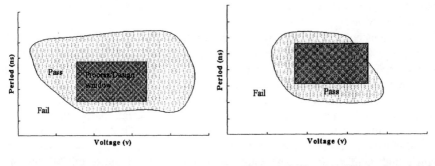

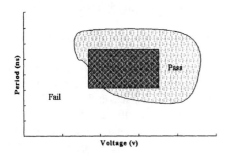

Figure 2.12. An ideal shmoo plot exhibits operating margin well beyond the required window defined by process and design (above left). A device with a high Vcc/low frequency problem is shown (above). A device with a low Vcc/high frequency problem is also shown (left).

The characterization of soft failure is best displayed by the use of the shmoo tool. The shmoo tool graphically depicts the passing and failing regions as parameters are varied together and against each other. In Figure 2.12, several shmoo plots of V_{DD} vs frequency are shown. The rectangle represents the design and process window for a device of a given technology. Marginality at a corner or edge of this window can be explained by common test and silicon failure mechanisms. For example, a device with a high voltage and low frequency problem (Figure 1.12b) might have test program timing issues, proximity design rule violations, or leakage (BVDSS). Similarly, a device with low voltage and high frequency problems (Figure 2.12c) might be failing due to an undersized transistor, a speed path, tester input/output timing, resistive interconnects, or overall transistor drive current.

Generally, temperature effects on device functionality vary inversely with the effects of V_{DD} voltage. Either increasing the voltage or decreasing the temperature will increase the speed of the device, while decreasing V_{DD} or increasing temperature will slow the device down. In most failure site isolation tools, it is difficult to control the temperature of the device, particularly to low temperatures, due to fixturing issues. Therefore, a temperature dependent functional failure should frequently be recharacterized as a V_{DD} dependent failure. This may include operating the device outside of the specified voltage range. In such cases, characterization of a correlation unit is important to verify the voltage range over which the device can normally be operated. Thus, it may be possible to reduce V_{DD} below the minimum specification limit to duplicate a high temperature failure or increase V_{DD} above the maximum to duplicate a low temperature failure. If the failing information from the datalog is the same in both situations, failure site isolation can be performed at room temperature.

2.4 MEMORY FUNCTIONAL FAILURES

Electrical characterization or Electrical Failure Analysis[17] (EFA) of memories is an important part of the overall failure analysis process. Increases in the structural complexity of memories has made their physical analysis and fault isolation more difficult and challenging. Electrical analysis can be used to characterize failures and provide an understanding of failing characteristics pertinent to each failure. During failure analysis, electrical characterization also strives to "fingerprint" new failure modes and mechanisms for recognition of future occurrences. Testing of a memory can be broken down into two components, array and periphery. The periphery is largely logic, and analysis of functional failures in the periphery is very similar to logic function failure analysis as discussed above. As with logic functional failure analysis, I_{DDQ} plays a critical role in physical failure site isolation. Partial and full array failures are also commonly analyzed as logic functional or parametric failures. However since the array is very highly structured, very detailed information can be obtained about array failures from electrical characterization[18]. This type of analysis relies on the built-in electrical functionality of the IC chip as a detector and sensor. The temperature and voltage dependence of failures are also useful since the voltage and temperature dependence of the detector, the device itself, is very well understood.

A defect in a DRAM memory cell array can manifest itself in a limited number of ways. These include (a) individual cell failure, (b) cell-to-wordline, (c) bitline-to-bitline, (d) wordline-to-wordline, (e) cell-to-cell, (f) cell-to-bitline, (g) wordline-to-bitline and (h) cell-to-substrate per Figure 2.13 below. A good understanding of the test programming language and the internal memory cell architecture are key factors

32 Electrical Characterization

in the successful implementation of EFA with test pattern algorithms to narrow down to the most probable failure mechanism.

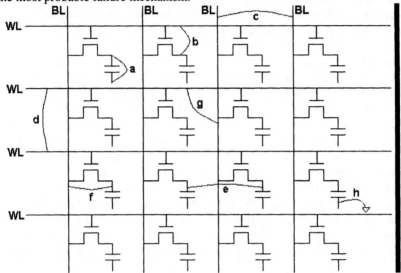

Figure 2.13. The possible failure manifestations in a DRAM array are illustrated.

2.4.1 EFA Test Program

Because of the highly structured array, EFA is able to provide very detailed information about a memory device. EFA programs are customized for failure analysis and differ from production test programs. They progressively check the functionality of the device so that at each stage it verifies the functionality of a portion of the circuit that can then be used to test other circuit areas. EFA programs are generally longer and include more algorithms than a production test program. A typical EFA program for DRAM's is summarized in Table 2.1. It is important to understand that the testing must be based on physical locations rather than address locations. It is an important consideration in many of the test algorithms such as adjacent row and column tests. EFA must take into account any redundant rows or columns that have been used during laser repair to understand the physical layout.

2.4.2 Single Bit Failures

Single bit failures form a large and very significant part of the failure distribution for memory devices. Single bit failures can represent stuck-at faults or an inability to maintain a particular state. A Write Immediate Read (WIR) test is used to test the most basic operation of a single cell, i.e. to store and read a "0" and a "1". A hard failure can usually be differentiated very readily from a marginal failure. For example, most DRAM designs incorporate a DFT feature to bias the cell top plate voltage "low" or "high". This allows cell dielectric failure to be differentiated from cell access gate failure. This test is also useful for detection of single cell stuck-at faults. A Single Bit Pause test is used to assess the ability of the capacitor to retain its stored charge. Because of the reverse-biased junction leakage current of the 1-T cell, the amount of stored charge (10^6 electrons) will decrease with time. A normal cell will exhibit reasonable refresh times of 64ms or greater but a defective one can range from <64ms to 0ms.

TYPE OF TEST	FAULT ISOLATION/PURPOSE
Internal Voltage monitor	Internal regulators functionality
VBB pump check	Leakage to substrate
Redundancy Check	Information on repaired row/column
V_{DD} shmoo	Single bit to row/column
DCR characterization	Margin for sensing
Write Immediate Read	How gross is the failure
Whole Array Disturb by Row/Column	Other row/column fail
Row/Column Pause	Worst bits on row/column
Adjacent Row/Column Disturb	Leakage from adjacent row/column
Open Address Pin characterization	Internal address malfunction
STIM level shmoo	STIM failure
Field Plate Program	Leakage paths
Failure Distribution	Package stress/characterization

Table 2.1. Details of a typical DRAM EFA Test Program are shown.

A Single Bit Disturb test (GALPAT = galloping pattern) is used to detect neighborhood pattern sensitive faults[19]. An array matrix of 8x8 cells is used to determine if a write operation on a nearby cell can change the contents of a base cell (cell under test) while the remaining cells and base cell contain a certain pattern. Each base cell must be read in state "0" and in state "1" for all possible changes in the neighborhood pattern. The neighborhood cells must be in physical proximity of the base cell because they are the most likely to influence the base cell and induce failure, rather than based on bit numbers. This test is also capable of detecting address faults, stuck-at faults, transition faults, and coupling faults.

The various possible leakage components within a DRAM cell are limited by the physical construction features of the cell. For single bit failures, differentiation between the various leakage paths can be easily achieved by varying the supply voltages. A simple Write-Immediate-Read (WIR) algorithm for a failing cell is developed to vary the cell top plate and device operating voltages. In the case of single bit failures, the failure site isolation is readily achieved down to the exact nature of the leakage path through EFA.

2.4.3 I_{DDQ} Testing

I_{DDQ} testing[20] has been used to detect defects which functional testing was unable to. A current mirror can be used to measure the current drawn by the device during a specific test cycle. This can be either the dynamic or the quiescent portion of the cycle. The tester is modified to read I_{DDQ} values into a bitmap. In this way, an I_{DDQ} threshold value can be set and I_{DDQ} values exceeding that threshold can be detected for every read through the memory array. In this way, individual bits, rows, and columns, which generate abnormal I_{DDQ} values, can be detected.

2.4.4 Other DRAM Characterization Features

DRAM's utilize a number of internal DC voltages such as V_{BB}, V_{PP}, V_{PERI}, V_{ARY}, V_{BLR}. These internal voltages are monitored through DFT tests on some devices or by bond pad probing on others. In addition, a different voltage can be forced on the power supply through the respective probe pads. This allows testing with different internal voltages on the chip. For example, forcing V_{BB} enables detailed Pause characteristics to be measured. Similarly, V_{PLT} can be forced to detect substrate leakage paths.

Curve tracing between V_{DD} and ground in a power down mode is useful for the assessing the gross functionality of a DRAM. As voltage is slowly ramped, current spikes indicate an internal state transition. This type of current spike can be used for subsequent global failure site isolation techniques.

The use of I-V analysis allows us to apply the right excitation techniques to other fault isolation techniques to enhance their effectiveness. The sequence in which tests are applied is critical if we are to interpret the results of the test program. We need to be aware of interaction effects especially when dealing with failures. The current limit set in tests must be carefully chosen and balanced between sensitivity and other effects. Electrical Failure Analysis is only one of the complement of tools used in the fault diagnosis process.

2.5 CHALLENGES OF ANALOG CIRCUIT FAULT ISOLATION AND ANALOG BUILDING BLOCKS

Electrical characterization of analog circuits for fault isolation poses unique challenges. Digital logic has several properties that make it well suited for testing and failure site isolation. Since the possible states of the system are limited to two values, 0 and 1, modeling failures is straightforward. This enables design for test (DFT) methodologies to be used that facilitate the diagnosis and isolation of failures. Low power (quiescent) states in logic devices also facilitate the use of I_{DDQ} test methodologies, which is useful as a global failure site isolation tool.

Analog circuits, on the other hand, have properties that make failure site isolation difficult. The outputs of analog circuits typically take on a continuous range of values and frequently have nonlinear transfer functions. This makes it difficult to model faults and facilitate DFT strategies which can assist in failure diagnosis and isolation[21-25]. In addition, analog circuits in general do not have power down states to support I_{DDQ} testing. Another significant challenge is that analog circuit failures are frequently associated with subtle wafer fab process variations, component matching sensitivity, or other design layout sensitivities. This results in failures that require extensive mechanical probing to isolate.

Analog circuits have a wide range of complexity but are generally composed of smaller building blocks such as op-amps, voltage references, and current sources. These building blocks can be combined, often with logic, to provide more complex analog and mixed signal devices. The following sections describe the characterization of several analog building blocks – voltage references and regulators, current sources, op-amps, and data conversion blocks.

2.5.1 Voltage References and Regulators

The objective of a voltage reference is to supply a known voltage that is stable over temperature and power supply variations. Figure 2.14 shows a block diagram of a bandgap reference circuit. It shows that the reference voltage is derived from the sum of a base-emitter voltage with the thermal voltage, kT. A base-emitter voltage has a negative temperature coefficient and the thermal voltage has a positive temperature coefficient, so that with a proper choice of constant, the output will have a zero temperature coefficient[26-29]. Failure modes associated with voltage reference generally fall into three categories; no output voltage, a stable output voltage that has the wrong value, or an output voltage that doesn't track correctly over temperature. Voltage references are feedback devices and will normally have two stable states, one at the correct output voltage and one at an incorrect output voltage, usually zero. Because of the existence of two stable states in voltage references, a start-up circuit is generally employed so that reference will power-up in the correct state.

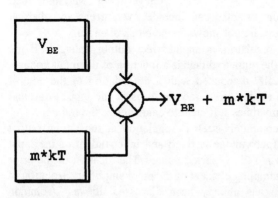

Figure 2.14. Block diagram of a bandgap voltage reference.

For voltage references with little or no output, the following characterization procedure can be used. If no leakage current is observed, the start-up circuitry is characterized to determine if it is operating properly. If it is not possible due to reference circuit architecture to isolate the start-up circuit for characterization, node voltages are measured and compared to a known good unit.

Voltage references that have an incorrect output voltage or do not track correctly over temperature are more difficult. Resistors on voltage references are typically trimmed at wafer probe in order to set the correct reference voltage. They are commonly trimmed using fuses or Zener diodes. The first step then would be to characterize the trims over temperature to assess their stability. Frequently a difference in the base-emitter voltage of two transistors (ΔV_{BE}) is set up in the reference by a ratio of resistor values. If one of these resistors is grossly off, say by an open contact, the reference voltage will be off also. Also, a high input offset voltage in the summer circuitry will also affect the reference voltage.

Voltage regulators use a voltage reference to produce a stable dc output voltage and maintain this voltage over a wide range of load currents and input voltages[30]. Figure 2.15 is a block diagram of a series pass regulator showing the major components. A sampling network monitors the output voltage. The sampled output voltage is compared to a stable reference voltage and produces an error signal. This error signal is used to control an element that converts the input voltage to the output voltage over variable load conditions. Additional protection circuitry can include overcurrent and thermal overload shutdown circuits but these are not shown in the block diagram.

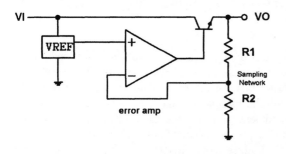

Figure 2.15. Block diagram of a series pass voltage regulator.

A typical failure for voltage regulators is an incorrect output voltage. If the device is not regulating, it means the output voltage is a function of the input voltage. This type of failure mode is usually associated with a degradation of the control element. For this type of failure mode, curve trace analysis of the input to ground and from the input to output is many times sufficient to characterize the failure.

A second functional failure commonly seen is regulation at the wrong output voltage. This means that the output voltage is fixed and independent of the input voltage, but the output voltage is not the expected value. The location of the error can be the voltage reference, the sampling element or the error amplifier. Isolation of failures in the voltage reference has already been discussed above. Common sampling elements consist of a resistor divider network (Figure 2.15) with the feedback voltage defined by a ratio of resistor values. Resistors are measured in order to isolate failures such as resistive contacts. The last potential source of error involves the errors associated with the error amp. These errors are frequently due to low gain, high input offset voltage, or common mode rejection ratio.

2.5.2 Current Sources

A current source is another basic analog circuit building block. As the name implies, a current source is used to generate a known current that is used to bias the various circuits on an integrated circuit. Current sources are also used as active loads for amplifiers to help to increase the voltage gain. Characterization of current sources generally consists of measuring the reference current. If this is incorrect, the mismatch between the transistors is characterized.

2.5.3 Op-amps

Op-amps are differential amplifiers that are widely used as stand alone circuits or blocks in larger circuits. Functional failures on op-amps include those with outputs do not respond to any input stimulus and are generally stuck at V_{DD} or ground and units that oscillate. Units that oscillate generally do so when the op-amp is configured as a unity gain amplifier.

For op-amps whose outputs do not respond to any input stimulus, the I_{CC} current with the device in the failing state is compared it to a known good device. A significantly lower I_{CC} reading may indicate an internal node is open and a significantly higher I_{CC} may indicate an internal node that is shorted. If I_{CC} variation is not effective, internal node probing (with comparisons to a good device) is required to isolate the failure.

Oscillation at unity gain in op-amps configured as an amplifier is primarily due to problems in the internal compensation network. The compensation network generally consists of a capacitor and a resistor. If the op-amp operates correctly when configured at higher gain, then a good characterization technique for this type of failure is to plot the gain and phase vs. frequency when configured with higher gain. The gain and phase can give clues to what could be wrong with the compensation network. Resistance value measurements and capacitor leakage measurements usually isolate the defective component.

There are several parametric failure modes that can occur in op-amps. These failure modes include input offset voltage, input bias current, or maximum power supply current tests. Input offset voltage and input bias current failures are the most common failure modes analyzed.

Input bias current failures can be a result of several failure mechanisms. Since the requirements for input leakage are extremely high, failures can occur due to very low current levels. Some of the leakage failures will be resolved by standard global failure site isolation techniques. For example, photon emission microscopy will generally isolate gate leakage on the input transistors of MOS op-amps. If global techniques are not effective, the elements of the input structure are separated using focused ion beam milling or laser cutting and probed to isolate a defective element. This is often required in the case of damage input protection structures. In bipolar op-amps, excess base current on the input transistors is the common cause of input bias current failures. This excess current can be a result of junction damage due to an

38 Electrical Characterization

overstress condition or it may be the extra base current required of a transistor with low gain.

For both the bipolar and MOS case, the input offset voltage is a function of the mismatch in the components that make up the input differential pair. The characterization of these fails will then consist of measuring the relative mismatches in the respective components.

2.5.4 Data Converters

Data converters are the interface between the analog and digital world. They consist of analog to digital converters (ADC) and digital to analog converters (DAC). ADCs take an analog signal and converts it into a discrete time digital signal that can be processed by logic circuits or digital signal processors (figure 2.16). DAC's do the reverse and take a digital signal and converts it back into a continuous time analog signal. Data converters are more complex than the analog blocks described so far, but their importance as a building block for more complex mixed signal devices warrants their inclusion here. There are many parameters associated with data converters and many different architectures from which to choose from. However, the specification they all have in common, and one of the failure modes for this class of part, is linearity. For ADCs, the digital code output should be a linear function of the applied analog voltage. Likewise for DACs, the output analog voltage should be a linear function of the input digital code. Characterization of linearity is simply a measurement of the transfer function of the given converter. Linearity failures can be caused by several defects. If an internal node is shorted, the unit will likely fail functionally as well as numerous parametric tests, including linearity. If the data converter has an internal voltage reference that is not working, the unit may fail the linearity test. In this case the characterization techniques described for voltage regulators should be used. If the data converter is functional and is still failing linearity, a likely failure mechanism is mismatched components (capacitors or resistors depending on the type of converter). For these types of failures, characterizing the matching characteristics of the critical components needs to be done.

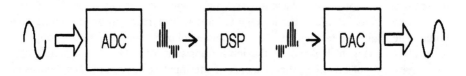

Figure 2.16. Block diagram shows that ADC and DAC link the analog and digital domain.

2.6 FUTURE CHALLENGES FOR CIRCUIT CHARACTERIZATION

Analog characterization is difficult and will become more difficult due to shrinking feature sizes, multiple metal layers, and increased device complexity. Analog characterization is heavily dependent on mechanical probing for precise DC measurements but shrinking device features and multiple metal layers will make mechanical probing extremely difficult without probe point creation. Also, devices will contain more analog blocks mixed with increasing levels of both logic and memory and the lack of well-defined testability methods will severely hamper analog analysis. Testability consists of both the controllability and observability of the IC and observability is particularly critical for failure analysis. The ability to identify and perform initial electrical characterization of failing analog blocks will be dependent on their testability. The success of characterization of analog circuits in large mixed signal devices will depend on the development of fault models, testing strategies, and software tools to extract critical electrical characterization information from the test data.

Voltage measurement-based test methods using stuck-at fault models has performed well to date for logic and memory. However, detection of faults[31,32] due to bridging and other physical defects that do not map directly onto stuck-at levels are expected to become more important. Probabilistic methods can be used to estimate whether this indeterminate logic value will be recognized and propagated as logic 0 or logic 1. Improvements in test modeling are expected to play a significant role in testing of memory and logic devices. These improvements are targeted at circuit controllability and observability. This process should also support more effective failure localization through testing. With shrinking geometries and thinner gate dielectric, bridging and gate-oxide shorts have become the more common defects and I_{DDQ} testing is currently the most effective methodology for detecting these defects and performing physical failure site isolation. Incorporating I_{DDQ} testing can bridge the current deficiency of performing logic testing alone by detecting defective devices that pass the functional test or logic testing. Hence by combining logic and I_{DDQ} testing, a very effective method to detect logical faults and physical defects in any circuit can be achieved.

Logic and memory DFT methodologies are driven by the opposing needs to improve test coverage and reduce test cost. Scan is likely to grow importance to increase test coverage of devices. Using BIST in logic to read the scan results is expected to help drive down the overall test costs. Similarly, BIST[33] for memory is expected to drive down test costs.

REFERENCES

1 Beall J, Wilson D. "Curve Tracer Applications and Hints for Failure Analysis." In Microelectronics Failure Analysis Desk Reference, 3rd Edition. Metals Park: ASM International, 1993.

2 Stabb D, Appleman D. Failure Analysis of Complex and High Pin Count Devices Using Computer Aided Electrical Characterization. Proceeding International Symposium for Testing and Failure Analysis, 1989, 261.

3   Appleman D, Wong F. Computerized Analysis and Comparison of IC Curve Trace Data and Other Device Characteristics. Proceeding International Symposium for Testing and Failure Analysis, 1990, 271.
4   Lycoudes N, Childers C. Semiconductor Instability Failure Mechanisms Review. IEEE Transactions of Reliability, R-29, (3), 1980, 237.
5   Sze S. M. Physics of Semiconductor Devices. John Wiley & Sons, 1981.
6   Efland T. Lateral DMOS Structure Development for Advanced Power Technologies. Texas Instruments Technical Journal, March – April 1994, 10.
7   Collins R. Excess Current Generation Due to Reverse Bias P-N Junction Stress. Applied Physics Letters. 1968, 13 (8), 264.
8   Collins DR. hFE Degradation Due to Reverse Bias Emitter-Base Junction Stress. IEEE Transactions on Electron Devices. ED-16 (4), 1969, 403.
9   McDonald BA. Avalanche Degradation of hFE. IEEE Transactions on Electron Devices, 1970, ED-17 (10), 871.
10   Schroder Dieter. Semiconductor Material and Device Characterization. John Wiley & Sons, 1990.
11   Amerasekera A,. Najm R. Failure Mechanisms in Semiconductor Devices. John Wiley & Sons, 1997.
12   Hawkins C, Soden J, Fritzemeier R, Horning L. Quiescent power supply current measurement for CMOS Ics. IEEE Trans. on Indus. Electron., 1989, 46 (2), 211.
13   Williams T, Kapur R, Mercer M, Dennard R, Maly W. IDDQ Testing for High Performance CMOS - The Next Ten Years. Proceedings European Design & Test Conference, 1996, 578.
14   Mentor Graphics. Understanding DFT basics, ASIC/IC Design-for-test Process Guide, V8_5_1, 1995, pp 2.1.
15   Butler KM, Johnson K, Platt J, Jones A, Saxena J. Automated Diagnosis in Testing and Failure Analysis. IEEE Design & Test, 1997, 14(3), 83.
16   Platt J, Butler KM, Venkataraman S, Hetherington G, Lorig G. Fault Diagnosis of the TMS320C80 (MVP) using FastScan. Proceeding International Symposium for Testing and Failure Analysis, 1996, 127.
17   Tan W, Chan A, Lam D, Swee YK. Electrical Failure Analysis In High Density DRAMs. IEEE International Workshop on Memory Technology, Design and Test Symposium, 1994, 26.
18   Lam D, Swee YK. Effective Test For Memories Based On Fault Models For Low PPM Defects. IEEE International Workshop on Memory Technology, Design and Test Symposium, 1993.
19   Van de Goor AJ. *Testing Semiconductor Memories Theory and Practice*. West Sussex, UK: John Wiley, 1991.
20   Lam D, Durai E, Swee YK. Implementation of IDDQ Testing for DRAMs. 2nd Memory Packaging and Test Conference, TI Singapore Internal Publication, July 1997.
21   Salama, Starzyk J, Bandler J. A Unified Decomposition Approach for Fault Location in Large Analog Circuits. IEEE Transactions on Circuits and Systems, 1984, CAS-31 (7), 609.
22   Milor L, Visvanathan V. Detection of Catastrophic Faults in Analog Integrated Circuits. IEEE Transactions on Computer-Aided Design, 1989, 8 (2), 114.
23   Hamida NB, Kaminska B. Multiple Fault Analog Circuit Testing by Sensitivity Analysis. Analog Integrated Circuits and Signal Processing, 1993, 4, 231.
24   Prasad VC, Babu NSC. On Minimal Set of Test Nodes for Fault Dictionary of Analog Circuit Fault Diagnosis. Journal of Electronic Testing: Theory and Applications, 1995, 7, 255.
25   Chao Y, Lin HJ, Milor L. Optimal Testing of VLSI Analog Circuits. IEEE Transactions on Computer-Aided Design of Integrated Circuits and Systems, 1997, 16 (1), 58.
26   Gray P, Meyer R. *Analysis and Design of Analog Integrated Circuits*. John Wiley & Sons, 1993.
27   Brokaw P. A Simple Three-Terminal IC Bandgap Reference. IEEE Journal of Solid-State Circuits, 1974, SC-9 (6), 388.
28   Michejda J, Kim S. A Precision CMOS Bandgap Reference. IEEE Journal of Solid-State Circuits, 1984, SC-19 (6), 1014.

29 Song, Gray P. A Precision Curvature-Compensated CMOS Bandgap Reference. IEEE Journal of Solid-State Circuits, 1983, SC-18 (6), 643.

30 Widlar R. New Developments in IC Voltage Regulators. IEEE Journal of Solid-State Circuits, 1971, SC-6 (1), 2.

31 Saxena J, Butler KM, Balachandran H, Lavo DB, Chess B, Larrabee T, Ferguson FJ. IE. Proceedings IEEE International Test Conference, 1998.

32 Lavo DB, Chess B, Larrabee T, Ferguson FJ, Saxena J, Butler KM. Bridging Fault Diagnosis is the Absence of Physical Information. Proceedings IEEE International Test Conference, 1997, 887.

33 Hii F, Powell T, Cline D. A Built-In Self Test Scheme for 256 Meg SDRAM. IEEE International Workshop on Memory Technology, Design and Test Symposium, 1996, 15.

3

PACKAGE ANALYSIS: SAM AND X-RAY

Thomas M. Moore
Cheryl D. Hartfield
Texas Instruments Incorporated

The detection of package related defects is an essential part of failure analysis. Non-destructive evaluations play a critical role in understanding the location and causes of assembly-related failures. At times, these techniques provide a complete understanding of a failure. At other times, destructive techniques must also be employed to understand the failure. The destructive techniques, typically decapsulation and cross-sectioning, are often mutually exclusive so that they can not be applied in series. In addition to guiding the selection of destructive procedures to complete the analysis, the non-destructive techniques provide an indication of areas to be exposed by decapsulation or cross-section. Scanning acoustic microscopy (SAM) and real-time x-ray radiography (RTX) are the primary techniques for the nondestructive imaging of the internal features of IC packages.

SAM is based on the focusing of an acoustic pulse at an interface within the package. Image contrast in the reflection (pulse-echo) mode is mostly due to the acoustic reflectivity at interfaces. The ideal spot size is diffraction limited and incident inspection wavelengths typically range from 6 - 150 μm in water (10 – 250 MHz). However, focusing aberrations, scattering and frequency dependent attenuation result in a practical spot size limit in the range of 50 – 250 μm depending on the type of package. Because sound is a matter wave, SAM is sensitive to cracks and delaminations within the package.

The wavelength of x-ray radiation is on the same order as interatomic spacings, so x-rays cannot be focused with a lens or reflected at internal interfaces (except at a glancing angle or as diffracted beams). RTX image contrast is based on the attenuation of an unfocused beam from a point source. The typical spatial resolution is better than 10 μm. RTX is sensitive to more strongly attenuating materials such as metal leads, eutectic die attach material and Au bond wires.

SAM and RTX are complementary rather than competitive in their capabilities. They provide different perspectives of the same package and each has its relative strengths and weaknesses. The instrumentation and capabilities of these techniques are presented, and practical examples in IC package inspection are discussed.

3.1 THE SCANNING ACOUSTIC MICROSCOPE

In pulse-echo SAM, the same transducer both transmits the incident pulses and receives the returning echoes as it is scanned in an image raster pattern. The sample and transducer are acoustically coupled by a water bath. Acoustic pulses are focused to a spot within the IC package (Figure 3.1) with a lens. Broad-band pulses are used to enable the differentiation of closely spaced layers. The echo signal is analyzed and characteristics of the signal, such as amplitude, phase and depth, are used to form images of internal structures and defects. The transducer is precisely scanned in a plane parallel to the surface of the package to produce an image at a fixed depth (C-scan image). Alternately, the equivalent of a nondestructive cross section image can be created by scanning the transducer in a line and displaying the reflected amplitude vs. depth (B-scan image). The pulse repetition rate is typically limited to 10 kHz due to decay of the reverberations that occur between the transducer and sample. Typical scan times for a package are less than 1 minute for a 256 by 256 pixel image.

The application of acoustic microscopy to IC package development was driven by the industry wide conversion to surface mount designs in the 1980's. Prior to this development, x-ray radiography (both film-based and real-time) was the primary method for nondestructive inspection of IC packages. IC packaging migrated from relatively small dies in robust, through-hole, typically dual in-line (DIP) packages, to larger, more complex devices in thin surface mount packages. Unlike wave solder assembly, surface mount assembly (vapor phase or infrared solder reflow) exposes the package body to the very rapid ramp to a higher temperature. This can result in the development of moisture/thermal-induced stresses sufficient to cause internal delamination and package cracking. The increased die size in many surface mount product results in higher stresses during temperature cycling. These stresses occur due to mismatch in the coefficient of thermal expansion (CTE) between the die and packaging materials. Reliability studies incorporating SAM inspection helped to clarify the moisture sensitivity issue, identifying delamination and not package cracking as the primary cause of electrical failure during temperature cycling[1-13]. These studies correlate the delamination revealed by nondestructive SAM inspection to the results of electrical testing and destructive physical analysis on both board assembly failures and reliability test failures.

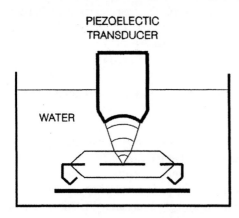

Figure 3.1. Inspection of IC packages with pulse-echo acoustic microscopy.

The more recent development of multi-layer substrate packages such as plastic ball grid arrays (BGAs) has resulted in the use of through-transmission SAM inspection of these substrates. The high attenuation in the organic materials and the fine layer spacing in these packages can make echo identification challenging in the pulse-echo mode of SAM inspection. Through transmission inspection makes use of a second transducer (receiver only) which is scanned with the transmitter. Through-transmission SAM enables rapid screening of organic BGA substrates at the sacrifice of the depth and phase information provided by pulse-echo SAM[14].

3.1.1 Image Contrast in the SAM

Image contrast in pulse-echo SAM inspection of IC packages is due primarily to the reflectivity of internal interfaces. The typical molded package provides a featureless front surface which is parallel to the desired image plane, and the interfaces of interest lie at approximately at the same depth. Thus, front surface morphology and absorption contrast due to path length differences do not dominate the image contrast. In through-transmission SAM, both interface reflection and attenuation losses contribute to the image contrast.

Sound is a matter wave. Matter waves depend on the vibration of molecules, atoms and electrons for propagation (in the same manner as heat conduction). Sound in the frequency range used for SAM has wavelengths similar to infrared radiation. The reflection and transmission of sound at interfaces can be described by geometric optics. However, there are some interesting differences between the behavior of light and sound at an interface. Unlike the refraction of light at the interface between isotropic media, sound produces two reflected and refracted waves in isotropic elastic solids (birefringence). The faster of these is a longitudinal wave in which the direction of particle motion is parallel to the direction of propagation. The other is a shear wave in which particle motion is perpendicular to the direction of the wave. Although shear wave imaging offers unique capabilities, this discussion will be limited to the primary imaging mode which uses longitudinal waves[15,16]. Also, in optical systems, light travels faster through the air than the lens. In acousto-optics, this situation is reversed. Sound typically travels faster through a solid lens than through the water couplant. So even though reflected and refracted wave directions in both systems are described by Snell's law, the sign of the radius of curvature of an acoustic lens will be opposite that of the corresponding optical lens (i.e.concave instead of convex).

At internal interfaces, a fraction of the incident acoustic energy is reflected and the remainder is transmitted. Figure 3.2 illustrates the example of a plane wave travelling through an ideal elastic solid (left) and impinging at normal incidence onto a planar interface with another ideal elastic solid

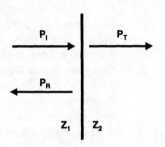

Figure 3.2. Reflection at an ideal interface is illustrated.

(right).

The incident plane wave has the sinusoidal acoustic pressure amplitude P_I and reflected and transmitted pressures amplitudes P_R and P_T, respectively. The boundary conditions at the interface state that the acoustic pressures and particle velocities must be equal in both materials. Thus, the frequency remains unchanged across the interface, and the reflectivity (R) and transmitivity (T) of the interface can be described by the equations below:

$$R = P_R / P_I = (Z_2 - Z_1)/(Z_2 + Z_1) \qquad T = P_T / P_I = 2 Z_2/(Z_2 + Z_1)$$

The acoustic impedances, Z_i, represent the ratios of the acoustic pressures to the particle velocities per unit area in each material. The acoustic impedances can be derived in the above example from the product of the density (ρ_i) and the speed of sound (v_i) in each layer[17-19].

$$Z_i = \rho_i \, v_i$$

Table 3.1 lists the acoustic parameters for some package materials. Note that at a bonded interface between plastic mold compound and Si, for example, R is positive and roughly 52%. However, at a delamination or package crack (which is represented by an interface between mold compound and air) 100% of the amplitude is ideally reflected (no transmission), and the phase of the reflected pulse is inverted relative to the incident pulse. Ideally both amplitude and phase inversion detection in the reflected signal can be used to identify internal delaminations.

Losses due to frequency dependent attenuation can be high in some packaging materials such as mold compounds (especially those with rubberized filler), adhesive die attach materials and organic substrates. Such attenuation losses in plastic packages and other factors such as interface roughness can often obscure the amplitude difference between reflections from delaminated and bonded interfaces. In these cases, phase inversion detection is an important tool to assist in the identification of delaminations and cracks in plastic packages.

The speed of sound in materials is typically less than 13 km/sec. This is roughly four orders of magnitude less than the speed of light. The time delay between echoes returning from a typical molded package can be easily measured electronically and images with three-dimensional information can be displayed. This is a unique advantage of a pulse-echo acoustic

Material	v (m/sec)	ρ (g/cc)	Z (10^5 g/cm^{2}sec)
Al_2O_3	10400	3.8	40
Cu	4400	8.9	39
Si	8430	2.4	20
Mold Comp.	~3500	1.8	6.3
Water	1480	1.0	1.5
Air	343	0.0012	0.00041

Table 3.1. Acoustic Parameters of Packaging Materials are shown.

technique and has been useful, for example, in determining the mechanism for package crack formation.

3.1.2 Image Resolution in the SAM

The spot size obtainable with a spherical lens (optical or acoustic) is limited by diffraction effects. Using the Rayleigh criterion, the lateral resolution[20-21], d, in a pulse-echo system is given by: $d = 1.02 \lambda F/D$, where λ is the acoustic wavelength, F is the focal length, and D is the diameter of the lens. The value of F/D ranges from 2 to 4 for typical transducers used for subsurface inspection in IC packages. So, practically speaking, the best resolution obtainable with these transducers is roughly two times the wavelength. At a center frequency of 75 MHz the wavelength in water is approximately 20 µm and the expected lateral resolution is roughly 40-80 µm. However, attenuation in a package acts as a low-pass filter and shifts the center of the pulse frequency distribution to a lower frequency. Focussing aberrations and scattering also degrade the spot size. This results in an increase in the minimum spot size. In practice, spot sizes are typically in the range of 50-250 µm, depending on the incident pulse frequency and package materials and thickness. Although transducer performance is the primary component affecting resolution, at pulse frequencies above ~100 MHz, it is the total integrated system performance that determines resolution, pulse width and efficiency. In addition to transducer performance, important system parameters include pulser and receiver bandwidth, cable effects and noise suppression.

Depth (or axial) resolution is important for distinguishing reflections from closely spaced layers within the package. Depth resolution in the time domain is determined by pulse duration as well as frequency. The inherent decay time for the transducer, focusing properties of the lens and frequency dependent attenuation all contribute to pulse duration. A typical pulse duration for a broad band transducer is 2 periods at the pulse center frequency. This effect creates what has been termed the "dead zone" below an interface. For example, for a duration of two periods at the center frequency, reflections from one interface may interfere with the reflection from an earlier interface and make detection difficult, especially if the signal amplitude is diminished by losses at the earlier interface. Real-time frequency domain analysis techniques may become practical for reducing this effect[23].

3.2 THE REAL-TIME X-RAY INSPECTION SYSTEM

The microfocus RTX system consists of three major components: the microfocus x-ray source, the sample and the image intensifier. Figure 3.4 shows a schematic of the geometry of the system. In the microfocus x-ray source, an electron beam (up to 360 keV in energy) is focused to a small spot on a replaceable target. The size of this focal spot determines the "geometric unsharpness" in the image. The sample holder allows for orientation of the IC package in 5 axes (x, y, z, rotation and tilt), and continuous adjustment of the magnification. The projected image of the package (shadowgraph) is detected with an image intensifier tube which consists of a fluorescent screen to convert the x-ray intensity image into light, and a camera or photomultiplier array to amplify the light image and convert it into digital format. Radiation shielding and safety interlocks are always an important aspect of x-ray radiography.

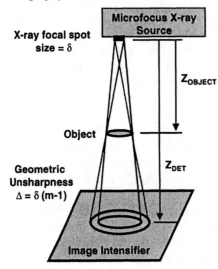

Figure 3.4: Microfocus RTX system schematic. Magnification: $m = Z_{DET} / Z_{OBJECT}$.

Real-time x-ray inspection systems are the successors of early x-ray fluoroscopes in which the transmitted x-ray image was projected onto a fluorescent screen. RTX systems with image intensifier tubes have been available since the 1950's. Dramatic improvements in the intensifier tubes in the 1970's expanded the application of RTX to the inspection of materials with higher attenuation (higher energy x-rays).

Current RTX systems offer high sensitivity (1% change in attenuation), a broad field of view, a spatial resolution better than 10 µm, flexible sample orientation, and sophisticated post-acquisition image processing. Real-time frame averaging improves the effective signal-to-noise ratio which results in images with better contrast and resolution. The post-processing options include automated defect recognition features.

RTX offers several advantages over film radiography. The ability to access the image almost immediately after acquisition is a major advantage. The flexibility to reorient the IC package and continuously change the magnification during inspection reduce the time required to obtain results and reveal defects that might otherwise go unnoticed. Image processing capabilities available for post-processing facilitate defect identification and enable RTX to compete with film radiography for image quality[23-26].

3.2.1 RTX Image Contrast and Resolution

Because the x-ray wavelength is similar to interatomic spacings, the x-ray photons interact with matter on an atomic scale. The primary interaction processes in the typical x-ray energy range for IC package inspection (<360 keV) are Rayleigh scattering, the photoelectric effect, and Compton scattering. Reflection and refraction at interfaces are replaced by diffraction at specific angles determined by the wavelength and the crystal structure of the diffracting material (Bragg's Law).

Thus, x-rays cannot be focused with conventional lenses. Reflection at a glancing angle (such as total internal reflection in a fiber) is possible but is very inefficient. Therefore, RTX image contrast is due primarily to differential x-ray attenuation in the sample. The attenuation is a function of the x-ray wavelength and the atomic number and thickness of the absorbing material. The reduction in x-ray intensity due to attenuation is a function of the linear absorption coefficient (μ) and the sample thickness (t) and can be expressed as:

$$I / I_0 = \exp(-\mu t)$$

The overall attenuation in a material is often expressed as the mass absorption coefficient (μ / ρ), where ρ is the material density. For example, denser materials such as metals (Cu lead frames and Au bond wires) have much higher mass absorption coefficients than less dense materials such as Si (dies), Al (bond wires) and plastic mold compound[23].

The spatial resolution in RTX images is primarily determined by the focal spot size of the microfocus source due to the geometric unsharpness factor described in Figure 3.4. Microfocus sources with reflection targets are available which provide a focal spot size of 5 µm. Thin targets (called transmission targets) offer even finer focal spot sizes (down to 1 µm).

3.3 APPLICATION EXAMPLES

Since SAM and RTX both provide non-destructive imaging of internal package structures, there can be confusion in choosing which of these two techniques is appropriate for identification of specific types of suspected defects. Each techniques has strengths or weakness in the detection of typical assembly related failures such as delaminations and package cracks, voids in flip chip solder bumps and die attach voids. The following examples illustrate practical applications of these techniques and illustrate the appropriateness of each technique for the identification of specific types of defects. They also illustrate the complementary capabilities of SAM and RTX in cases where there is not a specific failure mechanism suspected.

Figure 3.5 shows RTX and SAM images of the same plastic quad flat pack (PQFP) package. The contrast mechanisms at work in these two images are significantly different. The RTX image is a record of the relative attenuation of a broad beam of x-rays transmitted through the sample, while the SAM image shows the return echo amplitude of a focused acoustic spot that is mechanically translated to form an image. The acoustic echo signal has been gated at the appropriate time to selectively image the interfaces between the mold compound and the die and lead frame. The edges of the die are evident in the SAM image, and the bright corners of the die (arrow) are delaminations between the mold compound and the die surface. The RTX does not detect the delamination due to the relatively low attenuation of x-rays in air. The Au bond wires (25 µm) are not resolved in the SAM image and the overall spatial resolution is obviously better in the RTX image.

50 Package Analysis

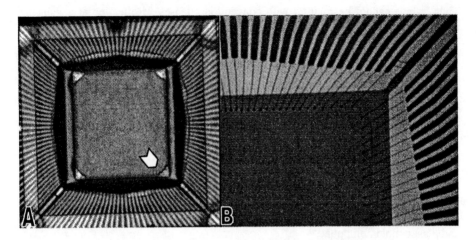

Figure 3.5. A plastic quad flat package (PQFP). A) SAM image (image area = 20 x 20 mm), and B) RTX image of the upper right corner of the same package.

The RTX image shows high attenuation at the Cu lead frame and Au bond wires. The low attenuation through Si makes the die relatively transparent in the RTX image. The area covered by the Ag-filled die attach is apparent and represents the area of coverage, but not necessarily the area bonded. As discussed in the following paragraph, SAM images can show the total bonded area at the interface between the die and the die attach material when the die attach reflection is selected for imaging.

Figure 3.6 shows an RTX image as well as two SAM images of the same thin quad flat pack (TQFP) package. In this case, the outline of the die is faint but discernable in the RTX image (arrows). The RTX image shows fairly good coverage of the die attach but it is heavily voided. Studies have documented that the RTX image of the die attach area remains the same during temperature cycling while periodic SAM images show progressive damage and the true percent area bonded[27]. In the TQFP SAM image shown here, the die attach is poorly bonded (Figure 3.6.C). The phase inversion image acquired by SAM (Figure 3.6.B) shows the presence of delaminations on the die pad and die surface. These are not detected by RTX. This is a good example of the complementary relationship between the two techniques: each is capable of providing information not revealed by the other technique.

The attenuation of X-rays in a sample can be so strong that limited information is available. Cracks in the ceramic portion of a microwave housing were detected initially by dye penetrant methods. Due to the 5 mm thick metal ring frame connected to the ceramic, the cracks were not detected by RTX inspection (area 1 in Figure 3.7). For the same reason, a void or delamination in the solder beneath the ring frame was also not detected by RTX (Figure 3.7, area 2). Areas 3, 4, and 5 in Figure 3.7 again display the strength of RTX for showing die attach coverage and the strength of SAM for showing true percent area bonded.

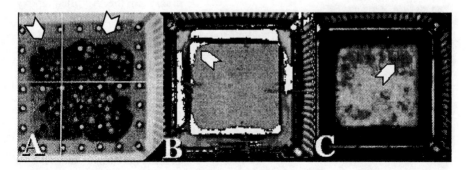

Figure 3.6. Images of a TQFP are shown: A) RTX image (arrow heads point to outline of die), B) SAM image at die surface (all white areas are delaminations), C) acoustic image at die attach (only darker regions (arrow) are bonded). (Image area = 9 x 9 mm)

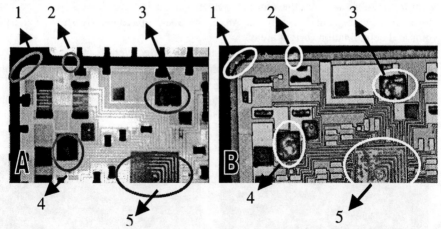

Figure 3.7. Housing comprised of a thick metal ring frame (5 mm) bonded to a ceramic substrate with solder. A) RTX image, B) SAM image: Five areas of interest are marked on each image for comparison. (Image area: 38 x 18 mm)

The fine layer spacing and high acoustic attenuation in multi-layer organic BGA substrates makes echo identification difficult. Through-transmission SAM inspection provides a method for rapidly screening these substrates for delaminations. Figure 3.8 shows the comparison of the RTX and through-transmission SAM images of a 272-pin plastic BGA package. The circled area is a delamination between the die and die attach material that was detected by SAM (delaminations appear dark in through-transmission and bright in pulse-echo SAM). The RTX image indicates cracks in the die attach probably caused by shrinkage during curing that were not resolvable with SAM. This apparent contradiction is explained by the fact that the

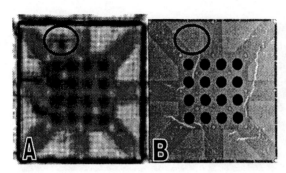

Figure 3.8. Images from the die area of a plastic 272-pin BGA. A) SAM image in through-transmission mode, and B) RTX image. (1 mm BGA ball pitch)

cracks in the Ag-filled epoxy die attach extend completely through the die attach layer while the die attach delamination detected by through-transmission SAM is a relatively thinner air layer that presents a negligable increase in total x-ray attenuation.

Flip chip underfill is an organic adhesive that has a relatively low x-ray mass absorption coefficient. The RTX image of an assembled flip chip device in Figure 3.9.B reveals patterned metal in the multi-layer substrate and the eutectic solder bumps but no defects are apparent. However, because sound is a matter wave, the SAM image clearly shows underfill voids and delamination (bright areas).

The RTX image in Figure 3.10 demonstrates the effectiveness of x-ray inspection for detecting small voids in eutectic flip chip bumps ~150 μm in diameter. Detecting these solder voids with SAM is difficult because of the resolution required (spatial and axial). This capability will be increasingly important as the size of flip chip bumps shrinks to accommodate finer pitch interconnect.

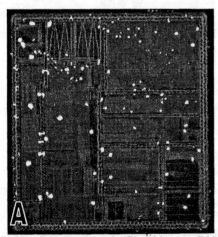 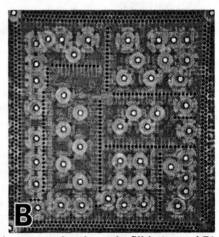

Figure 3.9. Images of a flip chip. A) SAM image gated at the underfill layer, and B) RTX image. (250 μm flip chip bump pitch).

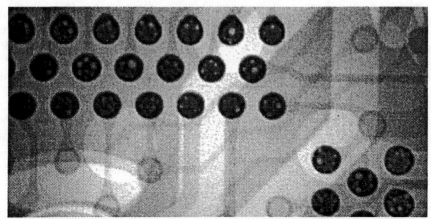

Figure 3.10. RTX image of eutectic solder flip chip bumps (250 μm pitch) readily shows small voids.

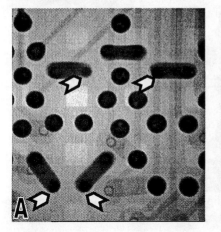 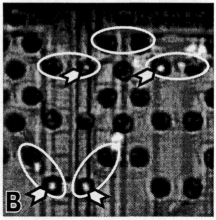

Figure 3.11. Images from a mounted but non-underfilled flip chip. A) RTX image, B) SAM image. Arrowheads show same positions in both images where an "anomolous" bright reflection is detected acoustically. Circles in the SAM image show where solder bridges occur.

The final example in this section is an analysis that required the capabilities of both SAM and RTX to understand the nature of the defect. The RTX image in Figure 3.11.A shows solder bridging in an organic substrate flip chip package that has not been underfilled. The SAM image (Figure 3.11.B) of the same location gated at the device surface shows bright contrast at specific solder bump locations. Bright reflections in SAM normally indicate cracks or delaminations which in this case would be an electrically open bump. However, in the plan view x-ray image (Figure 3.11.A) no evidence of open bumps are indicated.

54 Package Analysis

Figure 3.12 shows the RTX image of the same location as in Figure 3.11, but at an oblique angle. This image reveals that the solder bridges (Figure 3.11.A) are confined to the substrate. Furthermore, the tilted view RTX image identifies the bright spots in the SAM image (Figure 3.11.B) as pads on the die that are not connected to the corresponding pads on the substrate. Therefore, the plan view SAM image in Figure 3.11.B detects these "non-wets", whereas the plan view RTX inspection did not. However, the tilted view RTX image was required to understand the contrast in the SAM image. Without the discrepancy between the plan view RTX and SAM images, a high resolution RTX inspection on the tilted sample would not have been performed. Because the die was not underfilled in this case, the air gap between the substrate and the die prevented the SAM from detecting the solder bridging on the substrate.

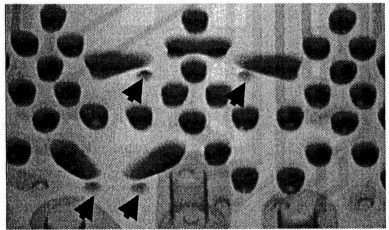

Figure 3.12. X-ray image shows the same flip chip as Fig. 13.11, taken at an oblique angle. Arrows correlate to positions marked by arrow heads in Fig. 13.10.

3.4 SUMMARY AND TRENDS IN NONDESTRUCTIVE INSPECTION

The examples presented in this section demonstrate the effectiveness of SAM and RTX for nondestructively detecting defects in packaged ICs. Both techniques provide images that are similar in many regards, but that have significant differences due to the contrast mechanisms at work. Image contrast in RTX images is based on the relative density of the materials in the package, while the pulse-echo SAM is sensitive to changes in the mechanical properties at interfaces. Although the attenuation of sound and x-rays as a function of depth can be described by very similar equations, it is often the case that materials that transmit sound readily are highly attenuating to x-rays and vice versa. For example, Cu heat slugs are highly attenuating to x-rays, but have very low attenuation to sound. In another comparison, the superior spatial resolution of RTX is needed to detect wire sweep and voids in flip chip solder bumps, but the sensitivity of the sound wave to changes in mechanical interface properties is required to image package cracks and delaminations. By comparing the information provided by these two techniques, one obtains a broader perspective of a packaging issue than could be obtained from either technique by itself.

The trend in pulse-echo SAM development is toward higher pulse center frequencies for better spatial and depth resolution. This is important for inspection of fine pitch flip chip interconnects, for example. However, frequency-dependent attenuation in the water path may set a practical upper limit on pulse frequency. Until this time, the analysis of the acoustic echo pulse has been done in the time domain. It is reasonable to anticipate the development of real-time frequency domain analysis capabilities to improve depth resolution for SAM inspection of advanced packages. Also, the image acquisition time in the SAM (less than 1 min.) can be greatly improved to the level of real-time by the development of two dimensional arrays for acoustic imaging[22].

RTX technology continues to produce improvements in source spot size and brightness. Replaceable thin transmission targets produce roughly a 5X reduction in source spot size by eliminating the large x-ray production volume that results from the interaction of the electron beam with a bulk target in the microfocus source. Continuing developments in post-acquisition image processing (post-processing) features extend the utility of RTX for defect detection.

Emerging nondestructive imaging techniques are being developed that may impact IC package inspection in the future. X-ray laminography uses a steerable x-ray source coupled with a rotating detector to produce cross section images within the sample. Because the practical spatial resolution is limited to roughly 125 µm, this technique is now seeing applications primarily in solder joint and printed circuit board inspection[28]. Imaging with far-infrared terahertz waves offers the potential for 100% inspection of packages without ionizing radiation or the need for a water couplant. Currently, the spatial resolution offered by THz imaging is roughly 250 µm and instrument cost and availability are still an issue[29].

ACKNOWLEDGEMENTS

The authors wish to acknowledge Jay Adams of CR Technology for the x-ray images of the flip chips.

REFERENCES

1   Lin, Blackshear E, Serisky P. Moisture Induced Package Cracking in Plastic Encapsulated Surface Mount Components During Solder Reflow Process. Proceedings International Reliability Physics Symposium, 1988, 83.

2   Moore TM. Identification of Package Defects in Plastic-Packaged Surface Mount IC's by Scanning Acoustic Microscopy. Proceedings International Symposium for Testing and Failure Analysis, 1989, 61.

3   Kuroki S, Oota K. High-Reliability Epoxy Molding Compound For Surface-Mounted Devices. Proceedings Electronic Components Conference, 1989, 885.

4   Nishimura A, Kawai S, Murakami G. Effect of Lead Frame Material on Plastic-Encapsulated IC Package Cracking Under Temperature Cycling. Proceedings Electronic Components Conference, 1989, 524.
5   Van der Wijk A, van Doorselaer K. Nondestructive Failure Analysis of ICs Using Scanning Acoustic Tomography (SCAT) and High Resolution X-ray Microscopy (HRXM). Proceedings International Symposium for Testing and Failure Analysis, 1989, 69.
6   Moore TM, McKenna R, Kelsall SJ. The Application of Scanning Acoustic Microscopy to Control Moisture/Thermal-Induced Package Defects. Proceedings International Symposium for Testing and Failure Analysis, 1990, 251.
7   Moore TM, McKenna R, Kelsall SJ. Correlation of Surface Mount Plastic Package Reliability Testing to Nondestructive Inspection by Scanning Acoustic Microscopy.   Proceedings International Reliability Physics Symposium, 1991, 160.
8   Moore, T.M., Kelsall, S.J., McKenna, R.G. "Moisture Sensitivity of Plastic Packages". In *Characterization of Electronic Packaging Materials*, Moore T.M., McKenna, R.G. ed. New York: Butterworth/Heinemann, 1993.
9   Van Gestel R, de Zeeuw K, van Gemert L, Bagerman E. Comparison of Delamination Effects Between Temperature-Cycling Test and Highly Accelerated Stress Test in Plastic-Packaged Devices. Proceedings International Symposium for Testing and Failure Analysis, 1992, 177.
10 Moore TM, Kelsall SJ. Impact of Delamination on Stress-Induced and Contamination-Related Failure in Surface Mount ICs. Proceedings International Reliability Physics Symposium, 1992, 169.
11 Shook RL. Moisture Sensitivity Characterization of Plastic Surface Mount Devices Using Scanning Acoustic Microscopy. Proceedings International Reliability Physics Symposium 1992, 157.
12 Van Doorselaer K, Moore TM, Tiziani R, Baelde W. Evaluation of Methods for Delamination Detection by Acoustic Microscopy in Plastic-Packaged Integrated Circuits. Proceedings International Symposium for Testing and Failure Analysis, 1992, 425.
13 Moore TM. The Impact of Acoustic Microscopy on the Development of Advanced IC Packages. Proceedings International Workshop on Semiconductor Characterization: Present Status and Future Needs, NIST, 1995, 202.
14 Moore TM, Hartfield CD. Through-Transmission Acoustic Inspection of Ball Grid Array (BGA) Packages. Proceedings International Symposium for Testing and Failure Analysis, 1997, 197.
15 Moore TM, Drescher-Krasicka E. Comparison Between Images of Damage and Internal Stress in IC Packages by Acoustic Microscopy. Proceedings International Workshop on Moisture in Microelectronics, NIST, 1996.
16 Drescher-Krasicka E, Willis JR. Mapping Stresses with Ultrasound. Nature, 1996, 7, 52.
17 Szilard, J. *Ultrasonic Testing*. New York: John Wiley and Sons, 1982.
18 Kinsler L.A., Frey A.R., Coppens A.B., Sanders J.V. *Fundamentals of Acoustics*, p. 125. New York: John Wiley and Sons, 1982.
19 Krautkramer, J. and Krautkramer, H. *Ultrasonic Testing of Materials*, 3rd ed. Springer-Verlag, 1983.
20 Briggs, A. *An Introduction to Scanning Acoustic Microscopy*. Royal Microscopy Society: Oxford University Press, 1985.
21 Briggs, A. *Acoustic Microscopy*. New York: Oxford University Press, 1992.
22 Moore TM, Hartfield CD. Trends in Nondestructive Imaging of IC Packages. Proceedings Characterization and Metrology for ULSI Technology, NIST, 1998.
23 Halmshaw, R. "Radiological Methods". In *Non-destructive Testing*, Honeycombe, R.W.K., Hancock, P., ed. London: Edward Arnold Publishers, 1987.
24 Bray, DE, Stanley, RK. "Radiographic Techniques in Nondestructive Evaluation" In *Nondestructive Evaluation: A Tool in Design, Manufacturing and Service*, Holman, J.P. ed. New York: McGraw-Hill, 1989.
25 ASM Committee on Radiographic Inspection, *Radiographic Inspection*. In Metals Handbook, (9th ed.), Vol. 17. Metals Park, OH: ASM International, 1989.
26 Colangelo, J. "Advanced Radiographic Techniques in Failure Analysis". In *Microelectronics Failure Analysis Desk Reference* (3rd ed.), Lee, T.W., Pabisetty, S.V., ed. Materials Park, Ohio: ASM International, 1994.

27 Moore TM, Frank K. Experience with Nondestructive Acoustic Inspection of Power IC's. Proceedings Electronic Components and Technology Conference, 1995, 305.

28 Adams, J. "X-ray Laminography". In *Characterization of Electronic Packaging Materials,* Moore ,T.M., McKenna, R.G. ed. New York: Butterworth/Manning, 1993.

29 Mittleman DM, Jacobsen RH, Nuss MC. T-Ray Imaging. IEEE Journal of Selected Topics in Quantum Electronics, 1996, 2, 679.

4

DIE EXPOSURE

Phuc D. Ngo
ST Microelectronics

If the physical cause of failure is on the die or elsewhere inside of the package, the first destructive failure analysis process is exposing the die and bonding. In general, these processes are intended to maintain the electrical characteristics of the device. This has become particularly critical as the number of pins has increased and the speed of the devices has increased. Maintaining an electrical signature of the failure for failure isolation is critical to success. Increases in device complexity and speed make it increasingly difficult to maintain electrical characteristics except through existing package connections. For cavity packages, exposure of the die typically consists in mechanically removing the lid, commonly called delidding. For plastic packages, removal of the mold compound covering the die and bond wires is called decapsulation. While historically exposing the top of the die has been desired, more recent emphasis has been placed on exposing the back of the die. This trend is being driven by the increase in flip-chip mounting. In addition the challenges posed by flip-chip technology, the increase number of layers of metallization has, in many cases, made it easier to isolate failures from the back of the die rather than the front. This consists of exposing the back of the die through removal of materials from the back of the chip. This can range from a heat sink to plastic material. After the bulk of the material is removal, the silicon surface is generally polished to facilitate IR light transmission.

In some cases where the failure mechanism is anticipated and chemical analysis is critical, mechanical opening without maintaining electrical connections may be preferred. Metallization corrosion failures are the most typical examples.

4.1 DELIDDING CAVITY PACKAGES

Since the die and wires in a cavity package are not connected to the lid, mechanical removal[1] of the lid will generally expose the die. Grinding away the lid is one approach, which has been successfully be used. In addition, the lid seal (material holding the lid to the package header) can be melted or cracked in order to remove the lid.

Grinding is used predominantly on ceramic packages with ceramic lids. In many ceramic lid packages, the lid seal also contains the pins as shown in Figure 4.1 for a C-DIP package. Cracking this seal can result in damage to the pins and loss of electrical connections. In these cases, grinding provides a more reliable method of maintaining connectivity. The easiest method is to use a diamond impregnated grinding wheel having coarse grit. Figure 4.1 shows a Cerdip prepared with this technique.

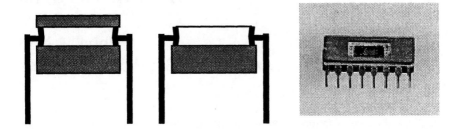

Figure 4.1. A cross-section drawing of a Cerdip package cavity before (left) and after grinding the lid off is shown. An optical photograph of a Cerdip after grinding on the lid is shown at the right.

Cracking the lid seal is also possible with most cavity packages. Most techniques employ either one or two knife edges. For ceramic seals, a pair of knife-edges is commonly used to crack the seal. Typically, the knife-edges are put together in an assembly with a pedestal which helps to maintain the device in a flat position between the knife-edges as shown in Figure 4.2. This type of assembly provides a very quick method for delidding. While most packages can be successfully cracked open without losing electrical connections, grinding is more reliable. Metal lid packages tend to be somewhat easier to manage since the external pins cannot be part of the lid seal as with C-DIP packages. There are generally two approaches used. The primary approach is to crack the lid off. One approach is to use a single knife-edge. The device is secured in position and the blade is placed in position to initiate a crack in the corner of the lid seal as shown in Figure 4.3. Typically a small hammer is used to drive the blade. Fixtures have also been developed to simplify this process, particularly for cavity down packages where the pins can interfere with placing the knife-edge. A second approach is to melt the lid seal, which is typically a eutectic alloy of gold and tin. Heat is applied

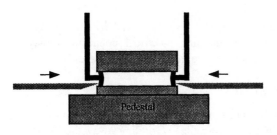

Figure 4.2. The diagram shows how a two blade system is used to crack open a Cerdip package.

to the lid until the alloy melts and the lid can be peeled up.

A few IC's, particularly analog devices, are packaged in cans. For these packages, the lid can be cut off. The tool used is somewhat akin to a copper-tubing cutter as shown in Figure 4.4. The device is rotated with the roller while forcing it against the knife-edge.

4.2 DECAPSULATION OF PLASTIC PACKAGES

Early acid decapsulation methods were extremely varied [2,3]. In some cases, the entire package was dissolved away, leaving behind the die and lead frame. A variation on this was to solder the lead frame to a paper clip in order to maintain the position of the bond wires. Obviously, neither of these techniques was very effective in maintaining electrical connections. With some variations, the industry migrated to a technique where a cavity was formed in the top of the package by milling. The device was placed on a heater block and a decapsulating acid (predominantly fuming nitric acid or fuming sulfuric acid) was dropped into the cavity. The acid was dumped and replenished until the die was exposed. This approach was reasonably successful for many years.

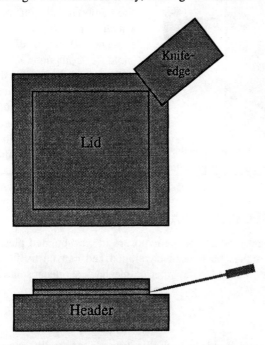

Figure 4.3. The procedure for cracking a metal lid seal is shown in top and side views.

Jet etch decapsulation[4] was developed largely to provide a more efficient method of decapsulation. A low-level vacuum is used to create a jet of hot decapsulating acid and to hold the device in place as shown in Figure 4.5. However, as mold compound has become more difficult to decapsulate, jet-etch has become a requirement for decapsulation. The jet etch provides several important advantages. The acid can be heated to a high temperature without excessively heating the device. In general, higher decapsulation temperatures are required for current devices. In addition, the time to decapsulate can be controlled by the temperature. If decapsulation time is too short, it is difficult to control the extent of etching. If the time is too long, acid tends to be adsorbed into the mold

62 Die Exposure

compound causing swelling and ultimately damage to the bonding. Fresh acid impinges on the device surface continuously until the die is exposed. This appears to prevent passivation of the etched surface. This passivation process can occur during the other approaches as the acid reacts and as clean-ups are performed for intermediate inspections.

One approach to facilitating the process of decapsulation is to reduce the amount of material which needs to be removed. If much of the mold compound is removed mechanically, the remaining mold compound can be removed much more quickly and efficiently. This applies equally well to other decapsulation techniques, which will be discussed.

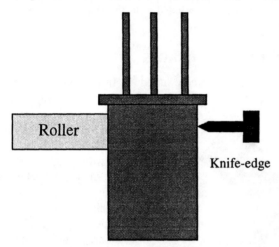

Figure 4.4. Delidding a metal can devices is normally performed on a tool similar to a pipe cutter.

4.3 ALTERNATIVE DECAPSULATION METHODS

Acid decapsulation is the dominant method for die exposure in wire-bonded plastic packages. However, other techniques have been developed and can be useful in some situations. Plasma, laser assisted, and thermo-mechanical decapsulation methods have been employed.

4.3.1 Plasma Decapsulation

Plasma decapsulation of mold compounds is possible using a primarily oxygen plasma. This attacks the organic components of the mold compound. However, the filler material (typically silicon dioxide to provide a thermal coefficient of expansion closer to silicon) is the primary component of the mold compound and it is not etched in the oxygen plasma. The typical ashing (a common term for plasma removal of organic materials) plasma is generated using oxygen mixed with less than ten percent CF_4. Hence the filler material must be removed separately. Plasma decapsulation is, therefore, typically an iterative process of ashing the polymer and mechanically removing the filler material. Since this is a time consuming process, removal of the bulk of the mold compound over the die can significantly decrease the time for decapsulation. The plasma reactions also generate significant heating of the device. This heating must be controlled in order to prevent damage to the device. On the other hand, the heating helps to speed up the plasma reaction and should not be totally eliminated. The plasma can also generate significant charging making it important to ground all of the pins.

Plasma decapsulation has not been widely used because it is time consuming to alternate between the filler and resin removal cycles. Alternative methods are preferred due to cycle time and throughput constraints.

4.3.2 Laser Decapsulation

As mold compounds have become more difficult to decapsulate, interest has grown in laser decapsulation. In the laser decapsulation process, the mold compound is precisely ablated. The benefits of laser milling applications include avoiding the stresses induced by mechanical drills and temperature cycling of lengthy wet etches. The laser milling can be performed such that a few mils of mold compound remain above the chip maintaining its electrical integrity. The remaining mold compound can be removed quickly with the standard wet etches. This technique becomes particularly applicable, as decapsulation of newer mold compound exceeds the capability of most jet etchers.

4.3.3 Thermomechanical Decapsulation

When failures are expected to be due to corrosion, analysis of the corrosion residue is a critical part of the analysis. Analysis of this residue provides a "fingerprint" of the source of ionic contamination, which contributed to the corrosion. Other types of failures where chemical analysis is critical can also be decapsulated thermomechanically[5,6,7], e.g. bond adhesion failures. Wet chemical decapsulation procedures typically remove corrosion products whose analysis is critical the failure analysis process. Thermomechanical techniques have been devised to break open the IC package without removing these corrosion products. Analysis of lead frame segments from a thermomechanically decapsulated device may also assist in identifying the migration path of contamination.

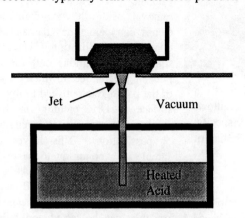

Figure 4.5. The basic element of jet etch decapsulation is a container of heated acid from which a jet is created by a vacuum.

The thermomechanical decapsulation techniques are extremely varied. They generally include one or more of three elements: some reduction of package size by grinding, heating the mold compound, and exertion

64 Die Exposure

of a mechanical force to crack the package or separate materials. One technique currently employed is to crack a heated device along the upper lead frame to mold compound interface. This approach is very similar to the technique described above for ceramic packages (Figure 4.2) except that a heated block replaces the pedestal. If the die surface is not exposed by the fracture, the top of the device is heated until the mold compound softens and the die can be lifted out with tweezers. In an earlier version, the backside was ground away to expose the die. The device was heated and the die lifted or pried out of the softened mold compound. With these approaches, the lead frame elements can be extracted from the mold compound as well as the die. A third approach has been to heat the device until it begins to smoke and twist the package with pliers. Innumerable other variations have been employed with varied success rates. Techniques may often be selected based on the sample size and required success rate. The primary disadvantage of this approach is the loss of electrical continuity, which makes it impractical for most decapsulation requirements.

4.3.4 Repackaging/Package Rework

Repackaging has been used for devices, which for one reason or another cannot be decapsulated or delidded with the die clearly exposed. A good example is current DRAM packaging where the leadframe is positioned on top of the die and the bond pads typically reside along the middle of the array. In order to achieve visibility of half of the die, the die is removed from the package and is wire bonded in a package (see Figure 4.6). Die removal with the standard wet etches can sometimes damage the bond pad metallization. This and some other situations call for the polishing technique in which the top of the package is polished to the die surface. Dry etching is then employed to remove poly-imide and or passivation. This leaves the sample in a condition to be tested and analyzed with probe cards or bonded into a package. Repackaging is also often needed for Tape Automated Bonding (TAB) assembly failures. Once assembled, these devices no longer have leads for testing. These failures must be re-packaged before they can be tested and failure analyzed.

Figure 4.6. EDO memory rebonded in different configurations (optical image) allows for analysis and characterization of specific circuits. Samples courtesy of Craig Salling of Texas Instruments Incorporated.

Many packages also require rework to prepare them for testing and analysis. Ball Grid Array (BGA) failures, for example, typically require "rebumping" or "reballing" after they has been removed from a board. Solder bumps on BGA packages can also be damaged during the decapsulation process. Malformed solder bumps are commonly reworked before the device is tested in a socket. The process involves removing and replacing all bumps in the array with interim steps of cleansing with solvents[9].

4.4 BACKSIDE PREPARATION FOR CHARACTERIZATION AND ANALYSIS

Several developments have created interest in accessing the device from the backside through the substrate. One is the growing application of flip-chip technology. Maintaining the ability to electrically stimulate an IC at speed is very difficult to achieve on high-speed IC's except through its normal interconnection environment. The second development is the rapid increase in the number of layers of metallization on a device. This blocks access to many areas of the device for conventional failure site isolation. Another case, for which backside analysis is important, is when design requirements cover the entire device with a power plane. Accessing the backside consist of three processes: removal of any heat sink or package material below the die, thinning and polishing the silicon and anti-reflective coating.

The removal of heat sinks and other materials from the back can be performed in a number of ways. The most common procedure is to use a milling machine with carbon steel bits to grind the material away. This technique sometimes employs a continual water rinse to reduce wear and tear on the bit and heating. In the case of ceramics, diamond encrusted bits are used. Computerized milling machines are commercially available for this application.

Once the backside of the die is exposed, a polishing compound is applied onto the silicon substrate. Dremel-like tools are sometimes used do the polishing. In most cases, a rotary polisher with some random motion is applied to maintain planarity. Iterative steps of inspecting the chip with an IR microscope would indicate when the active circuits are visible and suitable for analysis. Timed polishing can also be used since the removal rate is fairly consistent with the grit of the polishing compound. Most of the tools used for subsequent analysis benefit from removing as much silicon as possible. Removal down to a substrate thickness on the order of 100 microns has proven to be reasonably achievable. Techniques, which require greater thinning, are expected to be performed locally using techniques such as focused ion beam milling, chemically enhanced laser etching or mechanical methods.

An anti-reflective coating is typically required when the analytical technique employs an incident IR beam probe to extract data from the device. The coating minimizes beam reflection and refraction from the backside surface at the point of entry and in some cases, also aid with image resolution. The anti-reflective coating helps to maintain a high IR light intensity and improved signal to noise ratio.

In the event the die is in a cavity package or the device has been decapsulated, an epoxy must be used to fill the cavity to maintain the chip's stability during backside polishing (see Figure 4.7). This epoxy must have properties that will not alter the bonding as it cures. Backside preparations are typically more complex than topside.

66   Die Exposure

Figure 4.7. Optical images show the topside and backside of a Cerdip cavity which has been prepared for backside Emission Microscopy. The blurred topside view is a result of an epoxy used to fill the cavity.

4.5 FUTURE REQUIREMENTS

Access the die is a fundamental requirement for physical failure site isolation techniques. As flip chip devices become more common, improvements in the backside sample preparation will be required. In addition, as mold compound continue to become more difficult to remove, more innovative approaches such as laser decapsulation are expected to grow in importance.

REFERENCES

1 Bruenderman T. Opening Techniques for IC Ceramic Packages. Proceedings International Symposium for Testing and Failure Analysis, 1983, 190.
2 Lee TW. Chemical Decapsulation Revisited. Proceedings International Symposium for Testing and Failure Analysis, 1987, 113.
3 Wensink B. Improved Technique for Decapsulation of Epoxy-Packaged Semiconductor Devices and Microcircuits. Solid State Technology, October, 1979, 107.
4 Tang P. Techniques and New Etch Block Design to Enhance the Jet Etch Decapsulation. Proceedings International Symposium for Testing and Failure Analysis, 1985, 134.
5 Tomasi D. Failure Analysis Applications of Plasma. Proceedings International Symposium for Testing and Failure Analysis, 1987, 35.
6 Wagner LC, Boddicker S, Ngo P, Morgan D, Myers T. Failure Analysis of Plastic IC Package Integrity and Related Failure Mechanisms. New Technology in Electronic Packaging (ASM International, 1990), 353.
7 Wagner LC. Failure Analysis of Metallization Corrosion", ASM Conference on Electronic Packaging, 1987, 275.
8 Corum DL. Mechanical Decap Method for Plastic Devices. Proceedings International Symposium for Testing and Failure Analysis, 1984, 95.

5

GLOBAL FAILURE SITE ISOLATION: THERMAL TECHNIQUES

Daniel L. Barton
Sandia National Laboratories

The need for techniques that would produce high spatial and thermal resolution images of microelectronic devices has existed for many years. This became particularly true with the advent of multiple level metallization on integrated circuits (ICs). The addition of a second and subsequent levels of metallization significantly reduced defect observability and node access. Many defect types result in higher power supply currents, which generate heat during operation. This is due to the power dissipation associated with the excess current flow at the defect site. Systems to detect this power dissipation can be characterized by their sensitivity to thermal changes and spatial resolution. Infrared (IR) thermal techniques were the earliest available that calculated the temperature of an object from it's infrared emission. While IR techniques have excellent temperature range and resolution, they have a fundamental spatial resolution limitation. Liquid crystals have been used with great success since the mid-1960's. Liquid crystals provide a binary response, indicating if the hot area is above the crystal's transition temperature or not. Two factors have contributed to a recent reduction in the effectiveness of liquid crystals. Smaller feature sizes have made the spatial resolution of liquid crystal a factor. In addition, the reduction in power supply voltages and hence power dissipation have served to make the thermal sensitivity of liquid crystals an issue. The fluorescent microthermal imaging technique (FMI) was developed to overcome these issues. This chapter reviews the background material, operating principles, and image characteristics for these three thermal imaging techniques.

5.1 BLACKBODY RADIATION AND INFRARED THERMOGRAPHY

Blackbody radiation[1] physics describes the process by which all objects emit radiation to their surroundings. It is well known that heated objects emit radiation. The wavelength distribution of this radiation is dependent on the temperature of the object. For example, as objects are heated beyond the red hot temperature, approximately 700 °C, the amount of radiation being emitted in the visible range increases and the objects begin to turn orange, then yellow, and eventually to a bluish-white color. It is not readily apparent that most of the light being emitted by hot objects is beyond the spectral range of the human eye and in the infrared.

The connection between the peak radiation wavelength and temperature forms the basis for infrared (IR) temperature measurement. All blackbodies, objects that absorb all of the radiation incident upon them, that are at the same temperature emit radiation with the same spectral distribution, regardless of their composition. The spectral distribution of the radiation emitted by a blackbody is known as the spectral radiancy, $R_T(\nu)$, where ν is the frequency of the radiation. The quantity $R_T(\nu)$ is defined so that $R_T(\nu)d\nu$ is the energy emitted per unit time in the frequency range ν to $\nu + d\nu$ from a unit area on a surface at temperature T. The integral of the spectral radiancy over all frequencies is known as the radiancy, R_T, and is proportional to the temperature of the object to the fourth power. This relationship was first formulated in 1879 and is known as Stefan's Law. The relationship between wavelength peak and temperature is known as Wein's displacement law, which can be written,

$$\lambda_{MAX} = \frac{2.898 \cdot 10^{-3}}{T}$$

where T is the temperature in degrees Kelvin. The result of the equation is the peak wavelength, λ_{MAX}, in meters.

In most applications, the object under examination will not be a perfect blackbody. As such, the radiancy must be corrected for the emissivity of the material. The emissivity, e, is defined through Stephan's Law which relates the radiancy to temperature,

$$R(T) = e \cdot \sigma T^4.$$

By definition, the emissivity of a material is the ratio of the energy radiated by a given object to the energy radiated by a blackbody at the same temperature. The emissivities of many materials have been measured and are generally available.

In Figure 5.1, Planck's equation for blackbody radiation has been plotted to show that for objects at 2000 K (Figure 5.1(a)), there is only a small portion of the radiated energy in the visible portion of the spectrum. In fact, the majority of the information relating to the object's temperature is well into the infrared region of the electromagnetic spectrum. Figure 5.1(b) shows blackbody radiation spectra for objects between 250 K and 350 K, which is a more practical temperature range for semiconductor devices. Of particular interest in Figure 5.1(b) is the curve for 300 K, approximately room temperature. It is clear from this curve, that very little energy is emitted at wavelengths less than about 3 μm and most of the energy is emitted at wavelengths greater than 5 μm. In order to collect the infrared radiation information from objects near room temperature, a detector sensitive well into the infrared range is needed. The spatial resolution of this detector will be diffraction limited to something on the order of several microns[2].

It is evident from Figure 5.1(b) that the radiation emitted by the sample in the 3 μm to 12 μm range must be analyzed to measure typical device temperatures. IR thermal systems rely on directly sensing the emitted infrared radiation from objects to measure their temperature[3-6]. In general, there are several ways to accomplish this measurement. The simplest method is measure the radiance and, if the emissivity of the material is known, directly compute the temperature of the sample. Systems of this type use a very simple photovoltaic type detector that is sensitive in the IR wavelengths. Most IR thermography systems use one of two types of detectors, indium antimonide or mercury cadmium telluride. The most common are the cooled indium antimonide (InSb) detectors, which are sensitive in the wavelength range 1.5 μm to 5.5 μm. Mercury cadmium telluride (HgCdTe) detectors are sensitive over the range of 8 μm to 12 μm. Both detectors offer similar temperature sensitivities and ranges, but InSb operates at shorter wavelengths and should have somewhat better spatial resolution based on diffraction limits.

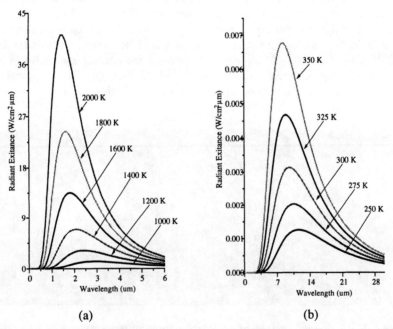

Figure 5.1. Spectral radiance of blackbodies at (a) 1000 K to 2000 and (b) 250 K to 350 K based on Planck's equation for blackbody radiation.

To measure the temperature of a sample, the radiance is measured first. Ideally, the radiance from a sample at a given temperature is related to the radiance that would be collected from a blackbody at the same temperature by the emissivity, or

$$R_T = e \cdot R_{TBB}$$

where R_T is the radiance from the sample, e is the sample's emissivity, and R_{TBB} is the radiance of a blackbody at the same temperature. If the sample's emissivity is known, its temperature can be calculated if the relationship between the radiance collected by a given system from a blackbody and its temperature is known. In order to increase the accuracy of the temperature measurement, the radiance that is reflected by the sample must also be accounted for as a background signal.

Lastly, the spectral response of the system must be considered. In general, the system response will not be constant over the spectral range on interest. The effect of spectral response can be characterized and incorporated into the blackbody radiance to temperature conversion algorithm.

Figure 5.2 shows example images of an integrated circuit test structure taken with an IR thermal imaging system. The structure is an electromigration test structure that has 4 µm wide aluminum lines separated by 4.5 µm. The lines are 3470 µm long yielding a total resistance of about 30 Ω. The images in Figure 5.2 show that IR systems can easily sense hot areas on integrated circuits at relatively low power densities but have difficulty resolving features less than about 15 µm. With the growing use of flip chip packaging technology, the need for backside localization of thermal features may create a resurgence in the use of IR thermal analysis for defect localization within integrated circuits. IR systems are particularly useful in areas such as multi-chip modules, circuit boards, and IC packaging issues where non-contact temperature measurements are essential and sub-micron spatial resolution is not needed.

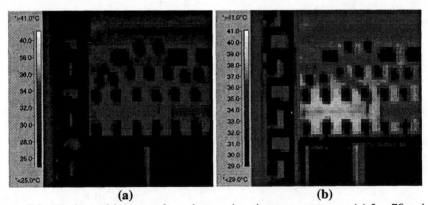

Figure 5.2. IR thermal images of an electromigration test structure. (a) I = 70 mA (actual surface temperature is 36.7°C), (b) I = 100 mA (actual surface temperature is 52.0°C.

5.2 LIQUID CRYSTALS

The history of liquid crystals[7-24] dates back more than a century. The first documented observance of a material that changed from a crystalline solid to an opaque liquid and then to a clear liquid as its temperature was increased was in 1888 by Friedrich Reinitzer. Reinitzer was synthesizing esters of cholesterol from plants and animals and noticed the state changes in these materials, which he called "double melting". Otto Lehmann, Germany's leading crystallographer at the time discovered the optical anisotropy in Reinitzer's esters, which lead him to create the terms, "fluid crystals" and "liquid crystals". Based on his observations, he was able to argue that the optical anisotropy, which he had observed, was due to elongated molecules in the opaque liquid. Daniel Vorländer started systematic research into the connection between molecular structure and the liquid crystalline state. In 1908 he established the rule that liquid crystal materials must have molecules with a linear shape. Between 1935 and about 1960, little significant research was done on liquid crystals. After 1960, interests in liquid crystal displays and thermography brought about a large resurgence in liquid crystal research.

A liquid crystalline state (of which there are many types) is one in which the material has some characteristics, which are associated with crystalline solids, and some which are associated with liquids, hence the name liquid crystals. The characteristics, which are like crystals, are usually associated with optical anisotropy while those of liquids relate to areas like molecular mobility and other fluid-like properties. Compared to isotropic liquids, liquid crystals have a much higher state of order while compared to solid crystals, they have a higher intermolecular and intramolecular mobility.

The materials used for failure analysis purposes have always been calamatic types. Calamatic liquid crystals have long, rod-shaped molecules as shown in Figure 5.3. Other types of molecules are discotic (disc shaped molecules) and sanidic (lath-like or board-like molecules). Most references use a shorthand notation to describe the temperature dependence of these materials. For example, 4,4'-azoxyanisole is described as "cr 118 N 135 is". It is a solid crystal at temperatures less than 118 °C, is a liquid crystal from 118 to 135 °C and becomes an isotropic liquid above 135 °C. The "N" in the notation refers to the type of liquid crystalline state that this material usually obtains, in this case nematic. Smectic types have a second level of order, which aligns the molecules in rows or layers. Cholesteric types were named after the cholesterols in which the phase was first identified and have a helical change in the alignment axis with depth. This difference gives cholesteric liquid crystals a unique ability to change colors by having reflection properties that are sensitive to wavelength. Figure 5.3 shows the differences in molecular order for the nematic, smectic, and cholesteric type liquid crystals along with a material in an isotropic state showing a much lower amount of order.

All of the calamatic types have a long rod-like molecular structure. The molecules are characterized by the presence of two or more linked aromatic structures (heterocyclic or alicyclic) with various organic groups attached to the rings. Changes in the various components of a given liquid crystal molecule affect the transition temperatures of each of the phases. Since most of the liquid crystals commonly used for hot spot detection on microelectronic devices are in the biphenyl family, examples from this family were chosen for Figure 5.4 to demonstrate the effects of molecule components on transition temperature.

The use of liquid crystals for thermal mapping was introduced in about 1963 and used cholesteric type materials. The technology was remarkable in that many thermally active objects, including integrated circuits, could be non-destructively tested in a fairly simple manner. Cholesteric liquid crystals have an interesting property of being able the change the wavelength of maximum scattering with temperature while they are in their cholesteric temperature range. Many references are available which describe the procedures for using cholesteric liquid crystals for thermal mapping [8-12].

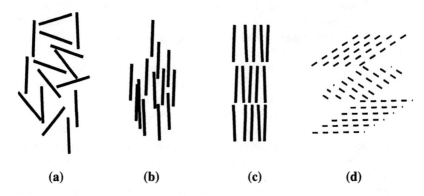

(a)   (b)   (c)   (d)

Figure 5.3. Ordering in Isotropic (a), Nematic (b), Smectic (c) and Cholesteric (d) liquid crystal states is compared.

Since nematic liquid crystals have become the materials of choice for thermal mapping of modern integrated circuits, their optical and thermal properties will be reviewed in more detail.

Two of the molecules shown in Figure 5.4 are the most common liquid crystals used in failure analysis. K-18 liquid crystal, which is chemically known as $C_{19}H_{21}N$ or 4-cyano-4'-n-hexyl-1, 1'biphenyl, has a state transition temperature from its nematic to isotropic state just above room temperature (Figure 5.4(c)). The low transition temperature of K-18 has undoubtedly lead to its popularity. K-21 (Figure 5.4(e)) has a transition temperature about 15 degrees higher than K-18 making it more suitable for integrated circuits with higher power consumption. The use of K-21 is more convenient than cooling the sample below K-18's transition temperature in most applications.

5.2.1 Optical Properties of Nematic Liquid Crystals

The use of nematic liquid crystals for failure analysis is based on their ability to give a detectable change when they are heated above a known temperature. Liquid crystals are well matched to failure analysis because their change with temperature is easy to detect with common optical equipment. Figure 5.5 shows the thermotropic optical birefringence of a nematic liquid crystal with a transition temperature of about 35 °C. Figure 5.5 illustrates the difference in refractive index between the ordinary ray (n_o) and the extraordinary ray (n_e) at temperatures below the transition temperature and total loss of birefringence above the transition temperature. The optical anisotropy of liquid crystals is due to the parallel order of the molecules while in the nematic state.

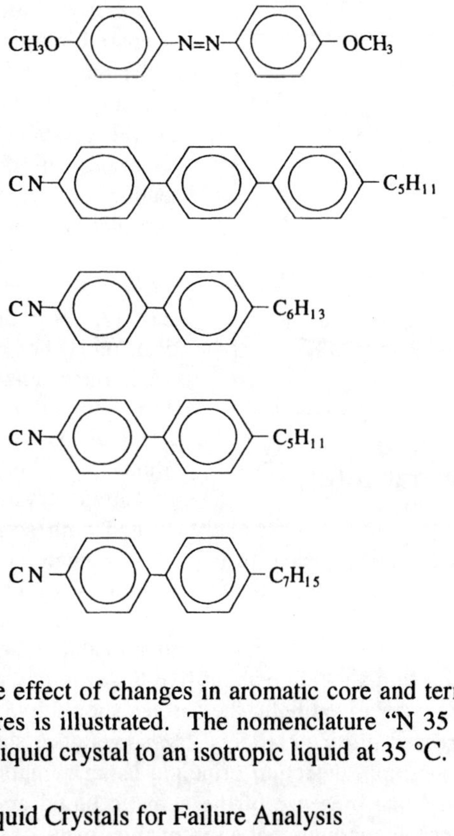

(a) CH₃O—⟨⟩—N=N—⟨⟩—OCH₃   cr 118 N 135 is

(b) CN—⟨⟩—⟨⟩—⟨⟩—C₅H₁₁   N 239 is

(c) CN—⟨⟩—⟨⟩—C₆H₁₃   N 35 is

(d) CN—⟨⟩—⟨⟩—C₅H₁₁   N 27 is

(e) CN—⟨⟩—⟨⟩—C₇H₁₅   N 42 is

Figure 5.4. The effect of changes in aromatic core and terminal groups on clearing point temperatures is illustrated. The nomenclature "N 35 is" indicates a transition from a nematic liquid crystal to an isotropic liquid at 35 °C.

5.2.2 Use of Liquid Crystals for Failure Analysis

The application of liquid crystals to integrated circuit failure analysis can be achieved in a number of different ways and with a variety of materials. The first applications used cholesteric liquid crystals[8-12]. In most of these applications, the circuit had to be covered with a thin, black material. The easiest coating was the carbon emitted by a common wax candle. The black layer is needed to allow the changes color to be observed under oblique illumination. The disadvantage of this method is that the opaque coating obscures the image of the circuit being tested. The references cited indicate spatial resolutions of 15 to 20 µm and temperature resolutions of 0.1 K at best.

Because of the spatial and thermal resolution limitations observed with cholesteric liquid crystals, nematic types[19-24] have replaced them. The main difference between the two applications is that cholesteric types generate color isotherms of the area being tested while the nematic types only indicate temperature changes by crossing the nematic to isotropic transition in the material. The ability to only detect that a given region on a circuit has heated the liquid crystal beyond the transition temperature is a limitation of the technique but is outweighed by the ease of use and sensitivity of this method.

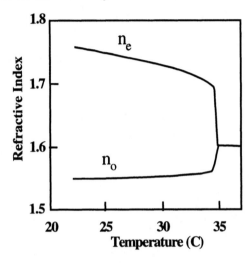

Figure 5.5. Thermotropic birefringence of a nematic liquid crystal with a clearing point of 35 °C is shown.

Infrared thermal imaging, as discussed in the previous section, is the only available non-contact thermal mapping technique. Liquid crystals and fluorescent microthermal imaging (see next section) both rely on the presence of a thermal sensing film applied to the sample being tested to provide an indication of thermal activity. In both techniques, the film must be thick enough to make the change detectable but thin enough to minimize the effect of the thermal mass of the film.

Liquid crystals are typically diluted with solvent (e.g. freon-TF or pentane) and applied to the surface with en eyedropper or syringe. The mixture is allowed to spread and the solvent allowed to evaporate. The resulting film should be about five to seven microns thick. When this film is viewed under un-polarized light, the surface should look slightly cloudy. When viewed through crossed linear polarizers, the appearance should be mottled. Figure 5.6 illustrates the hot spot detection principle using nematic liquid crystals. With the polarizers crossed, the presence of the nematic liquid crystal changes the polarization of the incident light enough that a reasonable image of the circuit can be made. Over the hot spot, the crystal goes into an isotropic phase where the polarization of the incident light is no longer altered. As a result, the "hot spot" areas appear dark in the microscope.

The entire equipment setup is illustrated in Figure 5.7. The coated sample is placed on a heated stage under a microscope which has two linear polarizers, one in front of the light source (referred to as the polarizer) and one in front of the camera or microscope eyepieces (typically called the analyzer). The most common technique is to heat the stage as close as possible to the transition temperature of the liquid crystal being used. Typically, the stage can be heated to within 0.1 K from the transition temperature with little difficulty. More elaborate procedures have been reported[20] to achieve stable temperatures much closer to the transition temperature. A variation is to heat the sample above the transition temperature and look for the last area to clear identifies the hottest portion of the circuit. By observing the liquid crystal while varying the temperature, the best thermal resolution can be achieved but only for a short period of time. Articles have been published[21] that describe the use a combination of stage heating and additional heating from the illumination source to achieve very high temperature sensitivity. Electrical stimulus for liquid crystal hot spot detection is typically a very low frequency AC signal. Pulsing current causes the hot spot to transition back and forth between states, giving a blinking effect. The amount of heating during each pulse and hence the size of the spot is controllable with the duty cycle of the AC signal.

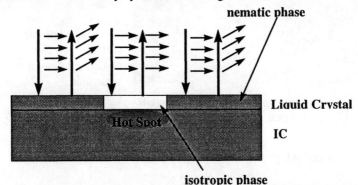

Figure 5.6. The use of nematic liquid crystals for failure analysis is illustrated.

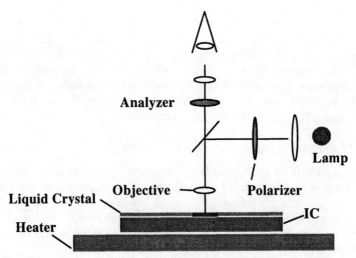

Figure 5.7. Equipment setup for liquid crystal hot spot detection is shown.

An example image of a hot spot detected with liquid crystal is shown in Figure 5.8. The image in Figure 5.8 was made using nematic liquid crystal K-18 while 100

76 Thermal Techniques

Figure 5.8 An example of hot spot detected using K-18 liquid crystal is shown.

mA of current was passed through the structure with the stage at room temperature. The electromigration test structure is the same as the one used in Figure 5.2. Since this structure is fabricated in the top level of metal for this technology, the heat generating features are the closest to the temperature sensing film making thermal features detectable on this structure at the lowest power density relative to other types of structures. The total area consumed by this metal pattern is 3470 μm in length by 4 μm in width or a total of 1.388×10^{-4} cm^2. Under a current load of 100 mA, this corresponds to a power density of only 2.161×10^3 W/cm^2 which is relatively low.

5.3 FLUORESCENT MICROTHERMAL IMAGING

The influences of lower power supply voltages, increased numbers of thin film layers, and smaller transistors have made liquid crystals less effective. Fluorescent Microthermal Imaging (FMI) provides a hot spot detection technique with greater thermal sensitivity and better spatial resolution than other, available techniques. The concept of using a film with a temperature dependent fluorescence quantum yield to generate high resolution thermal maps[25,26] of integrated circuits was first described in 1982-3. Subsequent research[27-29] has developed the understanding of the fundamental limitations of FMI and the operational procedures needed to insure maximum performance when applied to modern IC technologies.

5.3.1. Fluorescent Compounds

During the late 1950's and early 1960's, rare earth chelates were identified for use in lasers because of their well known fluorescence responses to UV or near-UV excitation sources. One of these compounds, EuTTA (europium thenoyltrifluoroacteonate)[30-33] has been the focal point for the fluorescent microthermal imaging technique. The chemical structure of EuTTA is shown in Figure 5.9.

EuTTA is not the only compound available for FMI. In fact, there are chelates of all of the rare earth elements, which include La, Sm, Eu, Gd, Tb, Dy, Tm, Yb, and Lu. The europium system was ultimately selected as the most suitable because of its temperature characteristics, emission/absorption characteristics, availability, and other qualities. There are several other europium compounds which might be suitable for FMI. Of the available compounds, EuTTA has the best temperature dependence of its fluorescence quantum yield for use near room temperature. For higher temperature measurements, another europium compound, perdeutero-(tris-6,6,7,7,8,8,8-heptafluoro-2,2-dimethyl-3,5-octandionato) europium (dEuFOD) may be used up to about 200 °C. dEuFOD has a much weaker temperature dependence both for fluorescence quantum yield and fluorescence lifetime than EuTTA[34].

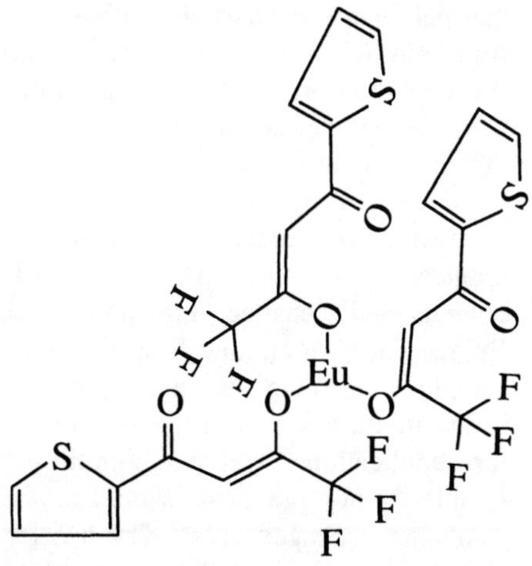

The ultraviolet radiation used to excite the EuTTA fluorescence does so through a series of intramolecular energy transfers. The TTA ligand absorbs the UV light then transfers the energy to the europium ion. Figure 5.10 shows the molar extinction coefficient (or the absorption spectra) versus wavelength for EuTTA in an ethanol solution. The broad absorption peak centered around 335 nm is where the TTA ligand absorbs energy. The lack of absorption for wavelengths much above 500 nm allows for a strong separation between the excitation source and the fluorescence emission. While several fluorescence lines are excited, the transition from the Eu^{3+} $^5D_0$ energy level to the $^7F_2$ level, as is shown in Figure 5.11, is the most efficient. This transition generates the bright fluorescence line at 612 nm that is used for FMI. Figure 5.12 shows the emission spectrum for crystalline EuTTA at 25 °C.

Figure 5.9. The chemical structure of EuTTA is shown.

For thermal imaging applications, an understanding of how the intensity of the fluorescent emission from the compound changes with temperature is required. Figure 5.13 shows the measured absolute quantum yield (fluorescence intensity) versus temperature for EuTTA in solution. For the FMI application, a calibration curve will need to be generated for each compound mixture that is used. The behavior of this compound is very predictable over a broad temperature range providing a simple way to calculate the temperature changes on a surface by imaging the changes in quantum yield.

The standard application technique for using EuTTA for FMI is to incorporate it into a PMMA (polymethylmethacrylate) matrix. A typical starting point is a solution consisting of 1.2 wt% EuTTA, 1.8 wt% PMMA, and 97 wt% MEK (methylethylketone). The MEK evaporates rapidly leaving the EuTTA/PMMA mixture on the sample[25]. Typically this mixture is spun on the sample and allowed to dry either at room temperature or in an oven for a few minutes. Ideally, the film

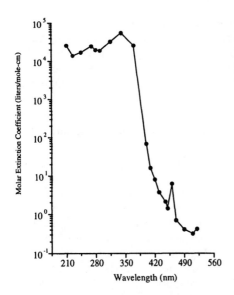

Figure 5.10. Absorption spectra for EuTTA (in solution)[30].

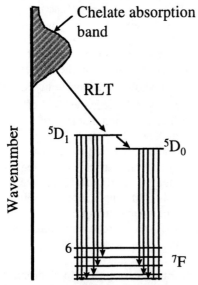

Figure 5.11. Energy transfer within he EuTTA molecule and the resulting Eu-ion transition corresponding to the 612 nm emission are illustrated.

should only be several optical absorption lengths thick, several hundred nanometers, which is thick enough that most of the UV light is absorbed, but thin enough that the thermal profile of the sample surface is not distorted. The image processing required to create a thermal image reduces the influence of film non-uniformity on image quality. As such, the film should be as uniform as possible but perfect uniformity of film thickness is not necessary.

The EuTTA/PMMA composition can be varied as needed for any specific application. Adjusting the EuTTA content will change the amount of fluorescent light emitted from the coated sample. Changing the amount of MEK in the mixture will thin the solution out for applications where spinning the sample is not practical, such as when analyzing packaged IC's. The use of a diluted mixture would allow a thin film to be deposited without spinning the IC. Usually, spinning a packaged IC will cause the mixture to accumulate around the ball bonds leaving a thicker film in these areas. The thick film is often not a problem, unless the signal input structures are the areas of interest. For these applications, a thinner mixture or a higher spin rate would be in order.

The advantage of using PMMA as a matrix is that it can easily be removed once the thermal analysis is completed. Rinsing the sample in acetone will dissolve the film in several minutes. The use of other polymers such dPMMA, (perdeutero-poly-methylmethacrylate) will provide stronger temperature dependence, but the additional cost of dPMMA is not justified. Other materials, such as spin-on glasses have been successfully used, but they make film removal more difficult.

In typical hot spot applications, detecting the area that is hotter than the background is all that is required. In some applications, an accurate determination of the absolute temperature is also required. Regardless of the film type used, accurate temperature measurements are possible with FMI. However, because of the differences in the logarithmic slope of the quantum efficiency versus temperature curves for different materials, an accurate film calibration should be done for each type of pre-mixed solution. The calibration curve can be easily obtained by using a hot/cold stage, a calibrated thermocouple, and the optical hardware and camera as used for FMI. The camera is used to collect the emission form the film in a given time period with the sample at a given temperature. Varying the hot/cold stage temperature over as large a range as possible yields the best results. Blank wafers are ideal for this process since, during the measurements, they will be close to the hot chuck temperature. The emission versus temperature data can easily be plotted from which a logarithmic slope can be calculated. Unless the composition of the mixture changes drastically, this measured slope need only be done once, especially if only relative temperature measurements are needed.

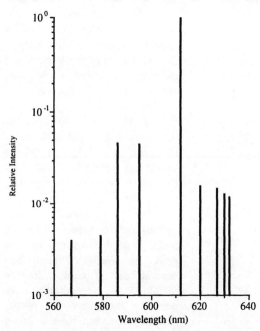

Figure 5.12. Emission spectra[30] of EuTTA (Crystalline) at 25 °C.

For EuTTA combined in a polymer or glass matrix, the fluorescence quantum efficiency will have similar temperature dependence, as the quantum efficiency is controlled by the intermolecular energy transfer process, but they will be different. Published values for the slope of the linear fit for EuTTA in a dPMMA matrix[26] are approximately -0.047 /°C. This number will be different for EuTTA in PMMA or spin-on glass, as well as for EuFOD in dPMMA. Whatever chemistry is chosen, a calibration curve must be made.

Extracting relative temperature changes is relatively simple. Collect an image without applying power to the sample (the cold image). Collect an image with power applied (the hot image). Divide the cold image by the hot image; take the natural log of the result and divide by the logarithmic slope. The result of this process is the relative temperature change at any pixel location[25,26]. The image processing is conveniently carried out with almost any image processing software. The only requirement is that the software uses the same number of bits per pixel as are received from the camera. Typically, cameras suitable for FMI will have 12 or 16-bits per pixel. If the software truncates the data to 8-bits, the thermal sensitivity will be adversely affected.

5.3.2 System Hardware

The most common FMI system design is shown in Figure 5.14. This has proven to be the easiest to assemble and best performing system design. There are essentially three main system components that are required for FMI. These components include a light source, a camera system, and an optical platform. This section discusses each of these areas and indicates differences between each of the possible choices.

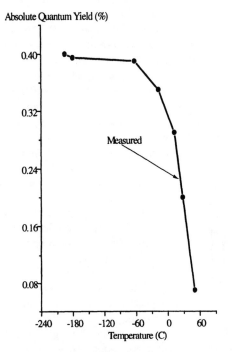

Figure 5.13. Measured absolute quantum yield for EuTTA[31].

The absorption spectra for EuTTA (Figure 5.10) indicates that ultraviolet light sources in the approximate range of 210 nm to 365 nm are the best match for EuTTA. The most common lamps for UV applications are mercury, xenon, and mercury/xenon types. Mercury bulbs have well-known spectral peaks in the UV range and several that extend into the visible range. Xenon lamps have peaks in the upper end of the visible range, but have a broad continuum of output that extends usefully into the UV. Mercury/Xenon lamps combine the two spectra to create a broad range, general-purpose light source. Because most systems are setup as shown in Figure 5.14, the broad wavelength output from the arc lamp filtered to the 365 - 390 nm band. The short wavelength cutoff comes from the optics in the microscope while a short wavelength pass filter controls the upper wavelength end. This is not the best match to the absorption spectrum of EuTTA. The absorption spectra shown in Figure 5.10 indicates that light at wavelengths longer than 365 nm will stimulate the fluorescence although with less efficiency than at shorter wavelengths. While this is true, sufficient fluorescence to perform FMI can be generated without the use of special optical UV lenses as discussed below.

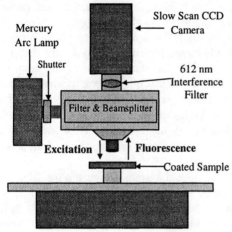

Figure 5.14. Implementation of FMI using an arc lamp and coincident illumination is shown.

The only requirement for the camera system is the ability to quantitatively measure the small changes in fluorescence intensity. The better the camera is at performing this task, the better the thermal resolution of the technique will be. Existing systems use slow-scan, CCD cameras. Slow-scan refers to the frame rate at which data is read out of the CCD array. Slow-scan cameras, since they do not adhere to TV standards, are designed to stare at a field of view and integrate for a variable length of time. For a situation where there is a very small amount of light being emitted, these cameras can stare at the field of view for several minutes to several hours or until the detector becomes saturated. In contrast, when using TV cameras for low light situations, it is necessary to grab and add video frames together, which also adds noise. Image averaging can be used to help remove noise, but it does not boost signal.

Dynamic range is also a significant factor in thermal resolution. Using a camera system with 8-bits of dynamic range limits system sensitivity to roughly a 0.4% change in quantum efficiency, or 1 part in 256. Slow-scan cameras are available with 12 to 16-bits of dynamic range and can thus image changes in intensity from 1 part in 4096 to 1 part in 65536. This translates into an order of magnitude gain in possible difference in temperature resolution. The spatial resolution can also be affected by the detector array size. Array sizes for slow-scan cameras range from 512 by 512 to more than 4096 by 4096 pixels.

Noise in collected images is another factor that limits temperature resolution. Slow-scan cameras are generally either peltier or liquid nitrogen cooled. Cooling reduces thermal generation of electron-hole pairs, which can fill up CCD charge wells with noise instead of image signal. Peltier, or liquid/peltier, cooled systems generally operate near -39 °C. At this temperature, only several electrons per second of noise are generated in an array with a charge well capacity of several hundred thousand electrons per pixel. At lower, liquid nitrogen temperatures, the noise is reduced to several electrons per hour. In these systems, the readout electronics also adds a small, but predictable amount of noise to the image, typically several electrons or tens of electrons per pixel.

In general, virtually any camera, which can yield an image in digital format, either directly or by frame grabbing, can be used for FMI. The use of a slow-scan CCD camera insures that the camera is not the limiting factor in system performance.

The key requirements for an optical platform are that it support the electrical stimulus required, have a port suitable for the UV illumination, and have a camera port. The two most common optical platforms in failure analysis are the probe stations and metallographic microscopes. Probe stations are constructed with electrical stimulus in mind but sacrifice some optical quality in exchange for extra long working distance lenses. Standard metallographic microscopes generally have superior optical systems, but suffer from having short working distance objectives and limited facilities for electrical biasing of the sample. Depending on the application, the optical platform that best suits the most frequent use of the system should be used.

The most important components in the optical system design are the filters and beamsplitter. The excitation filter should be a short wavelength pass (SP) filter with a 390 nm cutoff wavelength. The beamsplitter needs to reflect the UV light passed by the excitation filter but allow the visible fluorescence to pass with minimal attenuation.

Early FMI systems used UV grade fiber optic cable to obliquely illuminate the sample[25, 26]. This simplifies the optical system, but limits the amount of light that can be easily sent to the sample, especially when using high magnification, shorter working distance lenses. The use of a "through-the-lens" type of illumination reduces the problems of sample illumination, but adds the problems encountered with non-UV transparent optics found in most microscopes. Generally, standard optical components offer transparency for light with wavelengths greater than about 365 nm. UV grade optics are available because of the market for fluorescence and high spatial resolution work, but the components, such as lenses, tend to be very limited.

In order to minimize the amount of time that UV light from the excitation source is focused on the fluorescent film, a shutter needs to be placed in the beam path between the UV source and the sample under examination. The shutter is usually synchronized with the camera shutter. The UV light, which excites the fluorescence, also degrades the film generating non-thermal artifacts in the thermal images (see discussion on ultraviolet bleaching below). Although FMI can be performed without this shutter, its inclusion is a simple way to control one of the dominant noise sources. The final system component needed is an interference filter to filter out all of the fluorescence except the dominant line at 612 nm as shown in Figure 5.12. A filter with a bandwidth of about 2 - 4 nm is sufficient.

5.3.3 Photon Shot Noise and Signal Averaging

In absence of all other noise sources, the signal-to-noise ratio in high photon flux imaging applications is limited by photon shot noise, which follows a Poisson distribution. In order to obtain any meaningful thermal information, it is crucial that we can separate the thermal signals from the photon shot noise. Since photon shot noise is due to the quantum nature of light, there is no way to eliminate it totally; however, it is possible to reduce its effects through signal averaging.

5.3.4 Ultraviolet Film Bleaching

Inorganic-based films such as EuTTA are known to gradually lose their ability to fluoresce after exposure to UV light. The fluorescent intensity in these films decreases with increasing UV exposure time. This gradual loss in fluorescent intensity is known as bleaching. Since bleaching is unavoidable in EuTTA films, its behavior must be characterized to identify methods to minimize its effect. Figure 5.15 shows the bleaching behavior of three EuTTA/PMMA films of various dilutions to continuous exposure to UV light. In all three films, the fluorescent intensity decays rapidly initially and stabilizes after approximately 20 minutes of UV exposure. The diluted films, however, show a larger decrease in intensity before stabilization than the base mixture film.

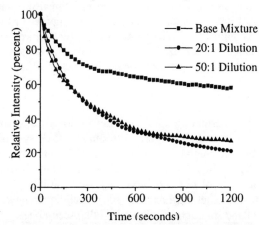

Figure 5.15. Changes in fluorescent yield versus UV exposure for three different mixture dilutions is shown.

In FMI, a thermal image is generated either by dividing the signals of a cold image by those of a subsequent hot image or vice versa. Ideally, the fluorescent

Figure 5.16. Overlay of FMI and reflected light images showing the location of a hot spot.

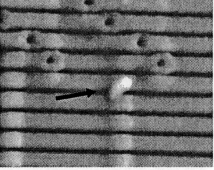

Figure 5.17. SEM image showing an embedded particle that produced an ohmic short between two adjacent metal lines in the failed SRAM.

signals of the hot and cold images should be identical in areas where no temperature changes occur. With bleaching, this ideal condition is not possible, resulting in the generation of unwanted, non-thermal signals.

5.3.5 Example Application

Figures 5.16 and 5.17 illustrate the application of the FMI technique to an integrated circuit failure. The example is a 1-megabit SRAM, which failed functional test and had an elevated I_{DDQ} of 5 mA at 5 V. Figure 5.16 shows a heat-generating area or hot spot that was located on this SRAM with FMI. The root cause of the failure was determined to be an embedded, stainless steel particle that shorted two adjacent, metal lines as shown in the scanning electron microscope image in Figure 5.17.

5.4 CONCLUSION

In this chapter, the subject of microthermal imaging for defect localization on integrated circuits has been reviewed in detail. The thermal resolution of IR systems is quite good for IC applications, but the spatial resolution has become an order of magnitude too large. The popularity of liquid crystals with failure analysts over several decades is attributable to its ease of use and flexibility with varied optical platforms. Finally, the fluorescent microthermal imaging (FMI) technique showed an improved combination of thermal and spatial resolution over other techniques.

In order to demonstrate the relative differences between these three thermal analysis techniques, images of the same structure at the same bias conditions are shown side by side in Figure 5.18. The infrared thermal image on top shows good thermal resolution and is a temperature map of the surface of the structure. The image also demonstrates the inherent spatial resolution limitation of this technique. It is the only non-contact thermal analysis technique and has potential for backside IC analysis.

The image in the middle of Figure 5.18 demonstrates the spatial resolution or liquid crystals along with the technique's limited ability to map surface temperatures. By the nature of the materials used for liquid crystal technique, it can only indicate the areas in the field of view that are above the nematic to isotropic state transition temperature. Liquid crystals remain a primary failure analysis tool because of their ease of use and excellent combination of thermal sensitivity and spatial resolution.

The final image on the bottom of Figure 5.18 was made using the FMI. This image demonstrates the unique combination of thermal and spatial resolution that FMI has. FMI provides only relative temperature maps that can be difficult to accurately convert to absolute temperatures. While FMI has been used with good success for data collection for thermal modeling of transistor technologies, which required absolute temperatures, this is really not the forte of FMI. FMI is intended to yield relative temperature maps of integrated circuit surfaces with very high resolution to quickly localize thermal defect signatures.

Overall, the three techniques reviewed in this chapter are and will remain the mainstay of thermal defect detection tools for the foreseeable future. Other, competing techniques which offer promises of either better thermal or spatial resolution are either still in their infancy or have other limitations which must be overcome before they can challenge the techniques described here.

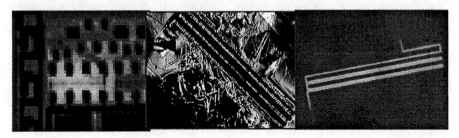

Figure 5.18. Comparison of IR thermal imaging (left), liquid crystals (middle), and FMI (right) on same test structure is shown.

ACKNOWLEDGMENTS

The author would like to thank Paiboon Tangyunyong at Sandia National Laboratories for his contributions to this research and Ronald P. Ruiz at the Jet Propulsion Laboratory for providing the infrared thermal images. This work was performed at Sandia National Laboratories and is supported by the U.S. Department of Energy under contract DE-AC04-94AL85000. Sandia is a multiprogram laboratory operated by Sandia Corporation, a Lockheed Martin Company, for the United States Department of Energy.

REFERENCES

1 Eisberg R, Resnick R. *Quantum Physics,* Chp. 1. New York: John Wiley and Sons, 1974.
2 Hecht E, Zajac A. *Optics*, Chp. 10. Reading, Massachusetts: Addison Wesley, 1974,.
3 Elliott CT, Day D, Wilson DJ. An Integrating Detector for Serial Scan Thermal Imaging. Infrared Physics, 1982, 22, 31.
4 Pote D, Thome G, Guthrie T. An Overview of Infrared Thermal Imaging Techniques in the Reliability and Failure Analysis of Power Transistors. Proceeding International Symposium for Testing and Failure Analysis, 1988, 63.
5 Zissis GJ. Infrared Technology Fundamentals. Optical Engineering, 1976, 15 (6), 484.
6 Burgraaf P. IR Imaging: Microscopy and Thermography. Semiconductor International, July, 1986, 58.
7 Stegmeyer H (ed). *Liquid Crystals*. New York: Springer, 1994.
8 Gray GW, McDonnell DG. Some Cholesteric Derivatives of S-(+)-4-(2'-Methylbutyl) Phenol. Mol. Cryst. Liq. Cryst., 1978, 48, 37.
9 Hiatt J. A Method of Detecting Hot Spots on Semiconductors Using Liquid Crystals, Proceeding International Reliability Physics Symposium, 1981, 130.
10 Dixon GD. Cholesteric Liquid Crystals in Non-Destructive Testing. Materials Evaluation, June 1977, 51.
11 Geol A, Gray A. Liquid Crystal Technique as a Failure Analysis Tool. Proceeding International Reliability Physics Symposium, 1980, 115.
12 Fergason JL. Liquid Crystals in Nondestructive Testing. Applied Optics, 1968, 7 (9), 1729.
13 Collings PJ, May JS. *Handbook of Liquid Crystals*. New York: Oxford University Press, 1997.

14 Khoo IC. *Liquid Crystals: Physical Properties and Nonlinear Phenomena.* New York: John Wiley and Sons, 1995.
15 Boller A, Cereghetti M, Schadt M, Scherrer H. Synthesis and some Physical Properties of Phenylpyrimidines. Mol. Cryst. Liq. Cryst., 1977, 42, 215.
16 Demus D, Richter L, Rürup CE, Sackmann H, Schubert H. Synthesis and Liquid Crystalline Properties of 4,4'-Disubstituted Biphenyls. Journal of Physics Colloque C1, 1975, 36, (supplement no. 3), C1-349.
17 Karamysheva LA, Kovshev EI, Pavluchenko AI, Roitman KV, Titov VV, Torgova SI, Grebenkin MF. New Heterocyclic Liquid Crystalline Compounds. Mol. Cryst. Liq. Cryst., 1981, 67, 241.
18 Laurlente M. Fergason JL. Liquid Crystals Plot the Hot Spots. Electronic Design, September 1967, Vol. 19, 71.
19 Hill GL, Agness JR. Practical Liquid Crystal Applications in Failure Analysis, Proceeding International Symposium for Testing and Failure Analysis, 1983, 73.
20 Ferrier S. Thermal and Optical Enhancements to Liquid Crystal Hot Spot Detection Methods. Proceeding International Symposium for Testing and Failure Analysis, 1997, 57.
21 Burgess D, Tan P. Improved Sensitivity for Hot Spot Detection Using Liquid Crystals. Proceeding International Reliability Physics Symposium, 1984, 119.
22 Burgess D, Trapp OD. Advanced Liquid Crystal for Improved Hot Spot Detection Sensitivity. Proceeding International Symposium for Testing and Failure Analysis, 1992, 341.
23 Flueren EM. A Very Sensitive, Simple Analysis Technique Using Nematic Liquid Crystals. Proceeding International Reliability Physics Symposium, 1983, 148.
24 Csendes A, Székely V, Rencz M. Thermal Mapping with Liquid Crystal Method. Microelectronic Engineering, 1996, vol. 31, 281.
25 Kolodner, Tyson JA. Microscopic fluorescent imaging of surface temperature profiles with 0.01 °C resolution. Applied Physics Letters, 1982, 40, 782.
26 Kolodner P, Tyson JA. Remote thermal imaging with 0.7 mm spatial resolution using temperature dependent fluorescent thin films. Applied Physics Letters, 1983, 42, 117.
27 Barton DL. Fluorescent microthermographic imaging. Proceedings International Symposium for Testing and Failure Analysis, 1994, 87.
28 Barton DL, Tangyunyong P. Fluorescent Microthermal Imaging - Theory and Methodology for Achieving High Thermal resolution Images. Microelectronic Engineering (Proceedings of the Fifth European Conference on Electron and Optical Beam Testing of Electronic Devices, August 27 - 30, 1995, Wuppertal, Germany)., February 1996, volume 31 (1-4), 271.
29 Tangyunyong P, Barton DL. Photon Statistics, Film Preparation, and Characterization in Fluorescent Microthermographic Imaging. Proceeding International Symposium for Testing and Failure Analysis, 1995, 79.
30 Winston H, Marsh OJ, Suzuki CK, Telk CL. Fluorescence of Europium Thenoylfrifluoroacetonate. I. Evaluation of Laser Threshold Parameters. J. Chem. Phys., 1963, 39 (2), 267.
31 Bhaumik M. Quenching and Temperature Dependence of Fluorescence in Rare-Earth Chelates. J. Chem. Phys., 1964, 40, 3711.
32 Crosby G, Whan R, Alire R. Intramolecular Energy Transfer in Rare Earth Chelates. Role of the Triplet State. J. Chem. Phys., 1961, 34, 743.
33 Bowen E, Sahu J. The Effect of Temperature on fluorescence of Solutions. J. Phys. Chem., 1959, 63, 4.
34 Kolodner P, Katzir A, Hartsough N. Noncontact surface temperature measurement during reactive-ion etching using fluorescent polymer films. Applied Physics Letters, April 1983, 42 (8), 15.

6

FAILURE SITE ISOLATION: PHOTON EMISSION MICROSCOPY OPTICAL/ELECTRON BEAM TECHNIQUES

Edward I. Cole Jr.
Daniel L. Barton
Sandia National Laboratories

Global failure analysis techniques are critical to keep pace with the increasing complexity of ICs. Global techniques provide methodologies for the isolation of failures without a detailed understanding of IC operation. Ideally, global analysis techniques should be easy to use, non-destructive, sensitive, and should provide high spatial resolution. In addition to the thermal methods described in the last chapter, other techniques generally fall into two categories: photon based and electron beam based. Photon probing of ICs has been and should continue to be a powerful approach to failure analysis. Because of silicon's relative transparency to infrared light, photon probing is particularly useful in cases where backside analysis provides the easiest access.

The most dominant global failure site isolation technique is photon emission microscopy. Localized light emitted by the IC during electrical operation is used in the analysis of the IC. Photon or light emission microscopy can be viewed as passive photon probing. Active photon probing takes advantage of a light beam's ability to interact with an IC and precisely localize the effects of this interaction. Two results of active photon probing, photocurrent generation by electron-hole pair production in semiconductors and heating, can both be used for defect localization. Photocurrents are used in the Optical Beam Induced Current (OBIC) and Light-Induced Voltage Alteration (LIVA) imaging techniques. Localized heating is employed in the Optical Beam Induced Resistance Change (OBIRCH), Thermally-Induced Voltage Alteration (TIVA), and Seebeck Effect Imaging (SEI) techniques. This chapter will discuss how the signals for both passive and active photon probing are produced and used in the diagnosis of IC failures.

Analogous to active photon probing, there are similar active electron beam probing methods that are powerful failure analysis tools. Electron Beam Induced Current (EBIC), Resistive Contrast Imaging (RCI), and Charge-Induced Voltage

Alteration (CIVA) are three important active electron probe techniques whose physics of signal generation and applications are also described.

6.1 PHOTON EMISSION MICROSCOPY

Photon (or light) Emission Microscopy (PEM) is a common failure analysis technique for semiconductor devices. The considerations involved in using photon emission to successfully analyze defects and failure mechanisms in CMOS ICs are well known[1,2]. IC analysis has typically been performed by collecting visible (390 - 770 nm) and some near infrared (770 - 1000 nm, with the NIR band defined as 770 - 1500 nm) wavelength photons emitted from transistors, *pn* junctions, and other photon-generating structures at or near the electrically-active, silicon surface. These photons are transmitted through the overlying, relatively transparent dielectric layers, passing between or scattered around the patterned, opaque metal interconnections. Detection of photons that emerge from around these overlying layers is referred to as front side PEM. Correspondingly, imaging light passing through the silicon substrate and emerging from the back is referred to as backside PEM. For both backside and front side PEM a reflected light image of the same field of view as the PEM image is acquired for registration of the light emitting areas. An infrared light microscope must used for backside, reflected light imaging. The reflected light images are normally combined in an overlay with the emission image into a single, processed image in order to facilitate localization.

There is increasing interest in backside PEM. This is driven by both increases in the use of flip-chip packaging and the increase in the number of layers of opaque metallization, which block the emissions. Backside analysis takes advantage of silicon's transmission of photons with energies less than its indirect bandgap energy, corresponding to wavelengths greater than 1.107 µm (for undoped silicon). It is commonly known that silicon becomes less transparent as dopants are added. Because of this, the heavily doped substrates, which are often used with newer technologies, attenuate NIR light emitted from the active circuits. To overcome this reduced transparency thinning of the silicon has become a standard requirement for backside PEM.

Different types of photon emission processes can be distinguished by their spectra. Radiative recombination emission is generally centered around 1.1 µm for silicon structures. Most defect-related emission is also strongest in the NIR range. System spectral characteristics are usually dependent upon the type of detector chosen. Detectors based on image intensifiers or CCD arrays have been the most commonly used. Most systems have very low spectral response to photons with wavelengths greater than 1 µm. The cameras used in these systems have spectral responses centered in the 400 - 900 nm range and can thus capture only a small portion of the emitted light. Detectors with extended NIR capability are essential for backside analysis and can increase the detection sensitivity for front side analysis.

6.1.1 Theory of Light Emission from Silicon ICs

In order to better understand the nature of the emission spectra from various processes in silicon devices, a general treatment of emission from indirect gap semiconductors is presented. General recombination processes in semiconductors can be broken down into two main categories: interband and intraband.

Radiative interband transitions occur when an electron in an excited state (occupying a state in the conduction band for example) recombines with a hole in a lower energy band (a vacant state in the valence band for example) to emit a photon. In indirect gap semiconductors such as silicon, this process must be accompanied by the emission or the absorption of a phonon as illustrated by the energy band diagram in Figure 6.1(a). Because of the relative lack of phonons for absorption at room temperature, this process is usually dominated by phonon emission. This means that, for indirect materials, there will be a significant amount of emission by a phonon (or a combination of phonons) at energies that are less than the bandgap or at slightly longer wavelengths. Additional spectral content can come from recombination involving impurity energy levels. Both donors (P at $E = E_C - 0.045$ eV and As at $E = E_C - 0.054$ eV) and acceptors (B at $E = E_V + 0.045$ eV) can further reduce the emitted photon energy and thus push the emission wavelength farther into the IR range as illustrated in Figure 6.1(b). Hot electrons (or holes) which have energy $kT_e = \Delta E$, represent the only major component by which light at energies significantly greater than the indirect bandgap can be emitted.

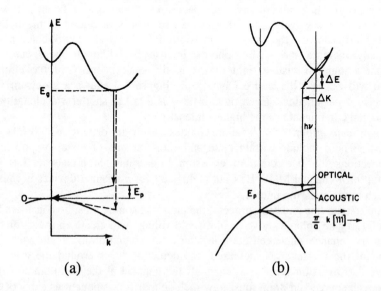

Figure 6.1. Energy band diagrams for indirect transitions (a) Phonon assisted transitions (b) Impurity and/or hot electron ($kT_e = \Delta E$) assisted[3] are shown.

Intraband recombination, as the name implies, occurs when an excited electron recombines with a vacant state in the same band. Usually, this process will not lead to a photon of significant energy for defect detection purposes unless the energy of the excited electron is more than 1 eV or so above the conduction (or below the valence) band edge. This type of emission is a component in the continuum commonly found in light emission spectra.

Photon emission from defects or abnormal operation of silicon microelectronic devices generally falls into only a few possible categories: forward or reverse biased *pn* junctions, transistors in saturation, latchup, and gate oxide breakdown.

Forward biased *pn* junctions have perhaps the easiest photon emission mechanism to understand. The mechanism is very similar to that of light emitting diodes (LEDs) and semiconductor lasers. The emission of these devices is generated a large population of electrons and holes in close physical proximity where they recombine. The recombination generates light with spectra centered around the bandgap. Because this is a low voltage emission mechanism, there will be few highly energetic carriers and the photon emission does not extend significantly toward shorter wavelengths. Figure 6.2 shows the emission spectra for a forward biased *pn* junctions[4].

Reverse biased *pn* junctions represent a different situation. When a small reverse bias is applied to a junction, the depletion region widens, producing a small current but a substantial electric field. As the reverse bias is increased, the probability that a highly excited electron will cross the junction increases. This will generate photons from the recombination of carriers whose energies can be significantly above the bandgap energy. The resulting emission spectrum will have a significant tail, which extends into visible wavelengths. If the reverse bias is increased to the point of avalanche breakdown, the added emission from charge multiplication will dramatically increase the visible photon intensity. Although there can be a significant amount of visible light emission, the peak intensity of the emission spectrum will be near the indirect bandgap. Figure 6.2 shows a spectrum for a reverse biased *pn* junction. Even though there is a tail at shorter wavelengths, the NIR emission component has the highest intensity.

A significant group of IC failure modes result in driving MOSFETs into saturation. These include certain open and short circuits of metal or polysilicon interconnections[1,5-7]. Detection of emission from saturated transistors has been successful for *n*-channel MOSFETs but is difficult for *p*-channel devices because of their much lower emission intensity.

To understand this emission process, the emission source must be considered. In a MOSFET, a gate voltage above the threshold voltage will create an inversion layer (a conductive channel) between the transistor's source and drain. If the voltage on the drain of the transistor is increased, the depletion layer around the drain will neutralize the inversion layer and pinch off the channel at the drain side. Charge flow between source and drain must now include a drift component as carriers cross the pinchoff region. Evidence has shown[1] that the light emission intensity is strongly coupled to the amount of substrate current. This indicates that the energetic carriers from the source or drain are radiatively recombining with majority carriers from the body of the MOSFET. Because this is similar to the reverse biased *pn* junction, the spectrum should be similar. Figure 6.3 shows a spectrum obtained from an *n*-channel MOSFET in saturation[5]. The similarity to the reverse bias curve in Figure 6.2 supports this hypothesis. The transistor in Figure 6.3 was biased with $V_{DS} = V_{GS}/2$, a situation known to maximize substrate current and photon emission[1]. Light emission from *p*-channel MOSFETs under the same conditions is more difficult to detect. This is due to ionization potential differences between electrons and holes, making emission from *p*-channel transistors significantly weaker. A third mechanism producing light emission is latchup. The classic model for latchup in CMOS ICs uses two parasitic bipolar transistors connected to form a silicon-controlled rectifier (SCR). When the SCR goes into its "ON" or low impedance state, all three of the SCR junctions are forward biased, placing the bipolar transistors in saturation. For these conditions, the expected emission spectrum will be of the same general shape shown in Figure 6.2 for a forward biased *pn* junction. The difference in this case is the relative area occupied by the parasitic SCR in comparison to a single *pn* junction on the same IC and the resulting amount of current required by the circuit during latchup.

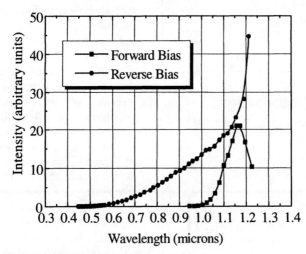

Figure 6.2. Measured spectra from forward and reverse biased silicon, *pn* junctions[4].

Perhaps the most difficult emission to explain is that of gate oxide shorts. This is a very common failure mechanism that is very difficult to locate without the use of light emission techniques. When gate oxide fails, a conductive path is created through the oxide. This connection forms a local alloy junction between the gate material, usually polycrystalline silicon, and the single crystal silicon in the channel region or in the source/drain region of the transistor. This situation would lead to an expected emission spectrum similar to those shown in Figure 6.2. Measured spectra (Figure 6.4) indicate that there is a strong component in the visible that decays in the NIR. This might indicate that the emission is thermionic, but the temperature needed to create a visible emission peak makes this unlikely. A simple calculation using Wei's displacement law gives a corresponding temperature for a thermionic emission

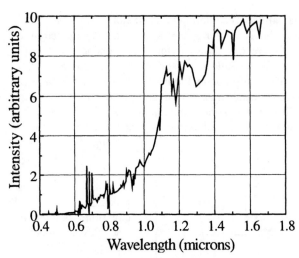

Figure 6.3. Measured spectrum from *n*-MOSFET saturation emission[8] is shown.

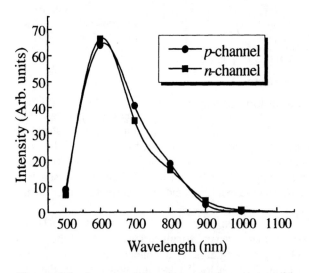

Figure 6.4. A comparison of spectra from gate oxide shorts in an 11 V technology is shown. The spectra were obtained using a series of bandpass filters at wavelengths from 400 to 1100 nm and corrected for both the filter response and the CCD quantum efficiency.

process of over 4800 K, which is clearly not physically possible for silicon ICs. Therefore the emission process must be dominated by electrons which have been excited to well above the conduction band edge (hot electrons) radiatively recombining with vacant states in the valence band.

Requirements for detecting photons are driven by several factors. Because of the recent popularity of flip-chip technologies and the rapid escalation in the number of metal interconnect layers which obscure the front-side view of transistors, the need for light emission technology for backside (through the silicon substrate) analysis has increased. This drives the need for a sensor with high sensitivity in the spectral region where silicon is transparent, the NIR. There are strong signals from all of the main light emitting defect types in the NIR. In fact, except for gate oxide failures, the strongest emission intensities are in the NIR. To meet these needs, both a sensitive detector and one with good quantum efficiency for wavelengths beyond 1 μm is needed.

6.1.2 PEM Detectors

Several types of detectors can be used for PEM. One of the most commonly used cameras is the liquid nitrogen-cooled, slow-scan camera with a thinned, back illuminated 512 by 512 pixel CCD array. The use of liquid nitrogen cooling effectively reduces the dark current noise to a negligible level, about two electrons per hour. The quantum efficiency (a measure of the detector's ability to collect light at a given wavelength) of a typical CCD array is shown in Figure 6.5. The peak in quantum efficiency is in the 600-700 nm wavelength band with a gradual drop in efficiency with increasing wavelength out to 1 μm. At 1 μm, the detector has a quantum efficiency of 17%. For longer wavelengths, the quantum efficiency drops quickly to only a few percent. The abrupt loss in quantum efficiency near the indirect bandgap of silicon is a direct consequence of the materials used to make the CCD array and suggests that silicon CCD detectors may not be well optimized for imaging emission processes in silicon ICs. The low efficiency for wavelengths to which silicon is transparent severely limits what can be detected from the backside. In most systems, the CCD camera is used to image the output of an image intensifier. Image intensifiers are multichannel plates which essentially amplify the light intensity. Unfortunately, they also amplify noise. Image intensifiers also facilitate gating or time resolution of emission. Image intensifiers exhibit spectral variations with the more recent version exhibiting somewhat better NIR performance.

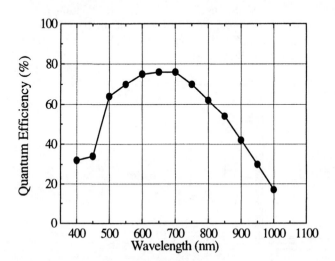

Figure 6.5. CCD array quantum efficiency is shown vs. wavelength[9].

A 256x256 pixel HgCdTe NIR array, NICMOS-3, was developed to image in the 800 - 2500 nm band on the Hubble Space Telescope[9]. This camera used a cooled, anti-reflection coated ZnSe window to provide good transmission between 500 nm and 5000 nm as well as low thermal emission. The usual read noise of the NICMOS-3 array is 35 electrons but modifications using special read techniques can reduce this noise to less than 15 electrons. Because the temperature of the array is maintained near 77 K, the dark current is negligible for integration times from 100 milliseconds to greater than 100 seconds. Thus, it is possible to measure photon fluxes at the array of less than 1 photon per second. Thermal blackbody radiation from the sample starts to become noticeable beyond 1400 nm and limits the sensitivity to non-thermal emission. In order to eliminate thermal information, a cold J-filter, which has a cutoff at 1400 nm and very low leakage at longer wavelengths can be placed in front of the array. A schematic diagram of this camera's main components is shown in Figure 6.6. PEM from the die backside is the application where infrared cameras significantly outperform CCD cameras. Figure 6.7 shows a typical example of backside PEM.

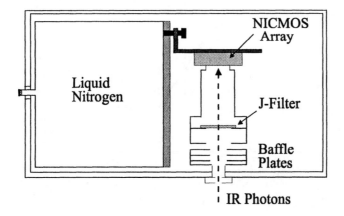

Figure 6.6. Layout of Infrared Laboratories camera setup as optimized for 1100 to 1400 nm.

Spectral analysis as shown above can be obtained by a number of methods. Band pass filters as described above are the most common method. For failure analysis applications bandpass filters normally are not used. Emission detection is normally done using the full spectral width of the detector to minimize integration time

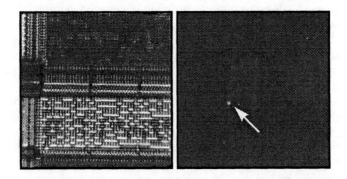

Figure 6.7. Infrared (1.1-1.4 μm) brightfield image (left) of gate oxide failure area and corresponding emission image (right) after an exposure time of 1.5 seconds.

6.2 ACTIVE PHOTON PROBING

Active photon probing takes advantage of the interactions of a scanned photon beam with an IC. In particular, for photon energies greater than the indirect band gap of silicon (≥ 1.1 eV or wavelength ≤ 1100 nm) electron-hole are generated pairs in the semiconductor. Normally the electron-hole pairs will randomly recombine and there is essentially no net effect. However, when electron-hole pairs are generated near the interface between differently doped regions in an unbiased IC, the charge carriers are separated by the built-in potential between areas with different Fermi levels as shown in Figure 6.8. Biasing an IC alters the Fermi levels and hence alters the magnitude of electron-hole pair separation or photocurrent. In contrast to passive photon probing techniques, the detector in active photon probing is the IC itself.

The photon beam source for active probing is usually a scanning optical microscope (SOM). The basic SOM consists of a focused light spot that is scanned over the sample in a raster fashion. The light source is typically a laser (LSM or Laser Scanning Microscope). While dependent upon the particulars of the optics used, spot sizes on the order of the laser wavelength used can be achieved. Spatial resolution is limited by carrier diffusion lengths and carrier lifetimes as well as spot size. By using different laser wavelengths and intensities, variations in the amount of photocurrent can be obtained and backside photon probing can be performed.

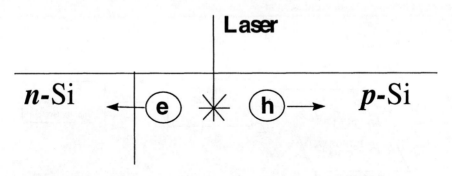

Figure 6.8. Photocurrent generation from photon produced electron hole pairs.

6.2.1 OBIC (Optical Beam Induced Current)

In OBIC, photocurrents are used directly to produce an image of Fermi level variations across an IC[11]. For typical operating conditions (front side examination with a 633 nm, 5 mW HeNe laser), the peak OBIC currents on an unbiased IC are on the order of tens of microamperes. The OBIC setup is shown schematically in Figure 6.9. Changes in the amount of electron-hole pair recombination current produce the contrast in an OBIC image. The resulting image displays the locations of buried diffusions on the sample as shown in Figure 6.10. In addition to junction location, variations in photocurrent can also result from defects associated with the junction such open interconnections, silicon crystal defects, and overstress damage. Any

96 Global Techniques

damage that alters the local Fermi levels on the IC will alter the amount of photocurrent produced. In addition to direct defect identification some semiconductor parameters such as minority carrier lifetimes, diffusion lengths, and surface recombination velocity can be determined by OBIC and the electron beam analog EBIC[12].

Electrical accessibility to the photocurrent generating site and interpretation of the OBIC signal limit the use of OBIC on ICs. Abnormal photocurrent production may be undetected if they are not readily observable via the IC's external interconnections. Electrical biasing can improve the observability of internal IC nodes, but large IC background currents can overwhelm the OBIC signals making them difficult to detect. Once acquired the OBIC signals of "good" vs "bad" ICs can be difficult to interpret, usually requiring a detailed knowledge of the IC design and comparison with a non-defective sample.

Photocurrent generation requires that photons reach the junctions of interest. Since metallization is opaque, signals can not be obtained in areas covered by metal. Backside examination of ICs has been well established and takes advantage of silicon's relative transparency to photons with energies just above the indirect silicon bandgap energy[10] (See Figure 6.11). Generation of OBIC signals from backside IR illumination requires that the photon wavelength be long enough to penetrate through the silicon substrate but short enough (have enough energy) to produce electron-hole pairs in the junction regions.

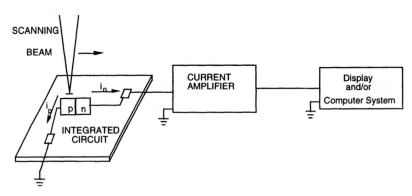

Figure 6.9. Typical OBIC setup is shown.

6.2.2 LIVA (Light-Induced Voltage Alteration)

LIVA images are produced by monitoring the voltage changes of the constant current power supply as the optical beam from the SOM is scanned across an IC[13]. Voltage changes occur when the electron-hole pair recombination current increases or decreases the power demands of the IC. LIVA, like OBIC, takes advantage of photon generated electron-hole pairs to yield information about IC defects and functionality.

In the "Voltage Alteration" mode, the IC acts as its own current-to-voltage amplifier, producing a much larger LIVA voltage signal than OBIC photocurrent signal. This is in part due to the difference in "scale" for IC voltage (mV-V) and current (na). Clearly the voltage signal is easier to measure.

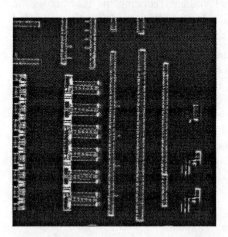 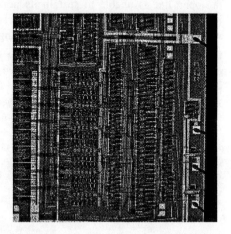

Figure 6.10. An OBIC image (left) shows the position of diffusions on an IC. A reflected light image (right) is shown for registration.

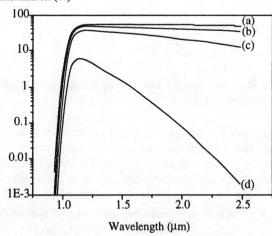

Figure 6.11. Percent transmission of light through 625 μm of p-doped silicon having doping concentrations of (a) 1.5×10^{16}, (b) 3.3×10^{17}, (c) 1.2×10^{18}, and (d) 7.3×10^{18} cm^{-3}.

Under identical illumination conditions, localized defects on ICs can generate LIVA signals 3 to 4 orders of magnitude greater than signals from non-defective ICs. This difference in LIVA signal depends upon the defect type, but two basic mechanisms can result in the large increase. First, the defect, because of its location in the IC amplifies the effects of normal photocurrents by altering the power demand of circuit elements connected to the defect region. Second, the defect region itself is a site of enhanced recombination compared to non-defective areas. Two types of defects illustrating the differences between these mechanisms are described below.

Junctions connected to open conductors amplify normal photocurrent effects to produce a larger LIVA signal (Figure 6.12). As the photo-produced charge flows, the open line is unable to sustain its normal voltage, reducing the bias across the junction. This can change the saturation condition of the transistor directly associated with the open-circuited junction, changing the IC's power demands. The voltage of the open-circuited conductor will be the same as the p^+ diffusion. Therefore any other transistors connected to the open conductor may change their saturation condition, further amplifying the LIVA signal. When photon injection ceases, the junction voltage will slowly recover to its initial equilibrium voltage, which is determined by weak coupling of the open conductor to neighboring conductors and transistors, parasitic leakage conditions, and tunneling across the open[14]. Of course, if the IC logic state is such that there is no potential difference across the open-circuited junction, there will be no LIVA signal. The use of LIVA on an IC with open conductors is illustrated in Figure 6.13. This example was performed from the backside using a 1064 nm, 1.2 W laser.

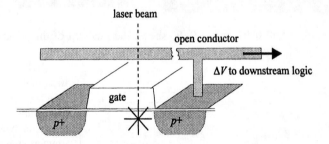

Figure 6.12. Outline diagram showing how localized photon injection can affect open-circuited junctions and downstream logic.

The other defect type with greatly enhanced LIVA signal is direct semiconductor damage such as overstress damage, crystal defects, and pinholes. Such semiconductor damage can cause a direct increase or decrease in recombination current. The changes in local Fermi levels caused by dopant redistribution and newly formed charge leakage paths will normally produce elevated I_{DDQ} with no illumination. Electron-hole pair generation and recombination due to illumination in the area of the defect will produce even greater amounts of "leakage current". When the supply current is held constant, the result will be a decrease in supply voltage and therefore an increased LIVA signal.

One final note about LIVA signal acquisition concerns the use of an ac coupled amplification system. An advantage of using ac coupled amplification is the mitigation of any dc offset signals that can complicate data acquisition. Another advantage is the use of "over-supply", increasing the effective bandwidth of the

LIVA system. Increased system bandwidth permits faster image acquisition without a loss in spatial resolution. The "over-supply" method involves increasing the supply current of the constant current source well above the maximum current needed to maintain the compliance voltage of the power supply. The compliance voltage prevents damage to the IC from overvoltage. If the constant current supply had infinite bandwidth, no LIVA signal would be produced under "over-supply" conditions. Because the current source does have bandwidth (response time) limitations, there will be a momentary reduction in supply voltage as the current source attempts to "keep up" with power demands. The LIVA signal can be produced by amplifying the momentary voltage reductions with the resultant system bandwidth being determined by the constant current source.

Figure 6.13. Backside IR LIVA (left) and reflected IR (right) images of a microcontroller with open contacts. The LIVA signal was detected while scanning the entire IC in a single image.

Logic State Mapping using both OBIC and LIVA can be acquired because photocurrent generation is dependent on the circuit bias. Figure 6.14 illustrates how the logic states of transistors can be identified. Figure 6.14b displays a LIVA difference image made from two images of the microprocessor in two different logic states. The field of view is the same as Figure 6.14a. The difference image was produced by a simple subtraction of two LIVA images, with the resultant image showing only those transistors that changed logic state.

6.2.3 IC Analysis Using Localized Heating from Photon Beams

Thus far we have discussed the production of electron-hole pairs and the subsequent photocurrents in semiconductors through photon interactions. Another active photon probing approach uses the heat from a photon beam to effect IC functionality. Through localized heating shorted, resistive, and open interconnections on ICs can be localized. Two methods for detecting shorts and resistive conductors are Optical Beam Induced Resistive Change (OBIRCH) and its constant current analog, Thermally-Induced Voltage Alteration (TIVA). Open conductors can be localized using Seebeck Effect Imaging (SEI). All of the thermal probing techniques can be applied from the front and backside of an IC by use of the proper optical wavelengths.

100 Global Techniques

To avoid photocurrent generation when performing localized thermal injection in these techniques, a laser wavelength with energy less than the silicon indirect bandgap is used. For the analysis described below, a 1340 nm laser is used. The relative transparency of silicon to infrared light (Figure x) permits backside analysis using this wavelength.

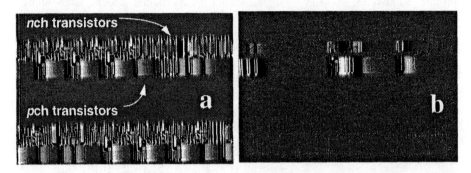

Figure 6.14. Two images showing (a) a LIVA logic map of cell rows for one state, (b) a LIVA difference image between two different states. The 5 mW, HeNe laser was used for the images.

6.2.3.1 OBIRCH and TIVA Imaging

Shorted conductors cause increased IC power consumption when the shorted conductors are at different electrical potentials, i.e. a short between V_{DD} and V_{SS}. The power consumption will depend upon the resistance of the short site and its location in the circuit. As a laser is scanned over an IC with a short circuit, laser heating changes the resistance of the short when it is illuminated, changing the IC power demand.

It has been found that thermally-induced power changes are usually greater for shorted signal lines than power busses. This results from signal line voltage fluctuations altering transistor gate voltages, producing the same amplification effect observed in LIVA. This resistance change with localized heating is the basis for the OBIRCH technique[15]. The change in IC power consumption is detected by an IC current change with constant voltage bias, yielding limited detection sensitivity in OBIRCH[15]. The same localized effect can be detected using a constant supply current biasing approach, which achieves greater sensitivity. This approach is known as constant current OBIRCH[16] or TIVA[17]. OBIRCH has also been used to localize defects with a high resistivity in conductors such as voids[16].

A front-side TIVA image of 3-level metal, 0.5 μm, 1 Mb SRAM is shown in Figure 6.15. The TIVA short site was identified as a stainless steel particle shorting two signal lines. Figure 6.16 is a higher magnification view of the short site. The particle causing the short and a small section of the shorted conductors can been seen in Figure 6.16(a). The particle is visible in Figure 6.16(b).

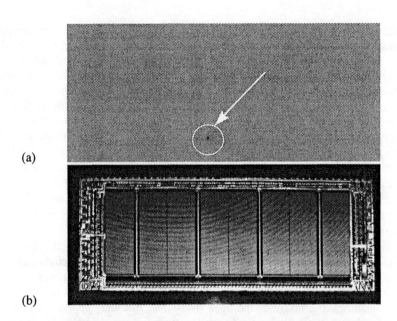

Figure 6.15. (a) Front-side TIVA image of a 1 Mb SRAM. The arrow indicates the short site. (b) Reflected light image for registration.

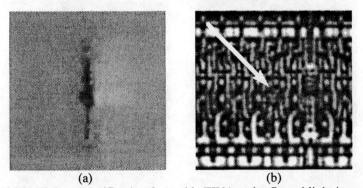

Figure 6.16. Higher magnification front-side TIVA and reflected light image pair of the short site in Figure 6.15. The shorting particle can be seen in Figure 6.16(b). The blurred appearance of Figure 6.16(b) demonstrates the spatial resolution limitations of imaging with the 1340 nm

6.2.3.2 Seebeck Effect Imaging (SEI)

Thermal gradients in conductors generate electrical potential gradients with typical values on the order of $\mu V/K$[18]. This is known as thermoelectric power or the Seebeck Effect[19]. The most common application of thermoelectric power is the thermocouple, which uses the difference in thermoelectric voltages of two different metals to measure temperature. For IC analysis, the effect has been demonstrated as a means to localize voiding in metal test patterns[20]. If an IC conductor is electrically intact and has no shorts, the potential gradient produced by localized heating is readily compensated for by the transistor or power bus electrically driving the conductor and essentially no signal is produced. However, if the conductor is electrically isolated from a driving transistor or power bus, the Seebeck Effect will change the potential of the conductor. This change in conductor potential will change the bias condition of transistors whose gates are connected to the electrically open conductor, changing the transistors' saturation condition and power dissipation. A Seebeck Effect Image[17] (SEI) of the changing IC power demands displays the location of electrically floating conductors. The use of constant current, voltage change measurement is critical to the success of this technique because of the small voltage alteration which occurs, typically on the order of μV. An example of a SEI image is shown in Figure 6.17.

The SEI signal is produced by heating the open conductor which occurs even if the scanned thermal probe is larger than the conductor is. The spatial resolution of SEI will be limited by the 1340 nm wavelength, however the detection sensitivity of SEI image contrast depends on the conductor being examined and not on the probe size. Therefore, the image of the open interconnection may appear larger than the actual conductor, but will effectively localize the open signal path.

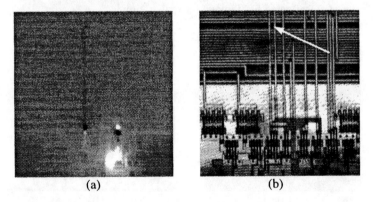

Figure 6.17. (a) Backside SEI and (b) reflected light images of an open conductor.

6.3 ACTIVE ELECTRON BEAM PROBING

The scanning electron microscope (SEM) is a necessary part of every failure analysis laboratory, commonly used for high-resolution images with a large depth of field. The SEM has become a powerful failure analysis tool because of the electron beam's ability to interact with an IC and precisely localize the effects of this interaction.

Several of the active photon probing tools developed for the SOM have electron beam analogs for defect localization. Electron Beam Induced Current (EBIC) is a corollary to OBIC that uses electron-hole pairs produced from primary electron beam/semiconductor interactions. Resistive Contrast Imaging (RCI), Charge-Induced Voltage Alteration (CIVA), and Low Energy CIVA (LECIVA) use direct and indirect charge injection to localize open interconnections on ICs.

The physics of signal generation, data acquisition techniques, and protocols for image collection are described below for each of these electron beam techniques.

6.3.1 Electron Beam Induced Current (EBIC)

EBIC[12] is perhaps the best direct comparison to its optical beam analog, OBIC. The images generated and the information produced can be almost identical, with the electron-hole pair current produced by primary electron interactions with the semiconductor under examination. Because of the similarities between EBIC and OBIC their differences will be highlighted here.

Differences in EBIC and OBIC occur because of the different ways in which the electron-hole pair current is produced. For OBIC, the photocurrent is produced if the semiconductor interacts with a visible photon from the front of the IC or an infrared photon of the appropriate wavelength from the backside. The primary electrons from a SEM will only be able to reach the silicon if the primary electron penetrates to that depth. The penetration of electrons can be controlled through the primary electron beam energy[21].

Since electron beam penetration for the typical range of beam energies (1-30 KeV) is only on the order of several microns, backside EBIC is not practical. For front-side analysis, EBIC has an advantage over OBIC in that the electron penetration is not dependent on the transparency. Since the penetration depth can be varied using the accelerating potential of the SEM, it is possible to tune EBIC images to a specific depth. This capability becomes more limited as the junction depth increases and the complexity of the structures above the junction increases.

Another major difference in EBIC and OBIC is the fact that a charged beam is used to create electron-hole pairs. This means that the sample can become charged through electron injection. Charging of insulators leads to trapped, fixed charge, which can result in parasitic MOS transistors or even electrostatic discharge damage.

The charge deposited in the gate oxide (fixed oxide charge) and the occupation of oxide/silicon interface trap sites will change the threshold voltage[22] of MOS transistors. Depending upon the primary electron dose to the gate oxide, the effects can range from an increase in IC leakage current to total functional failure of the transistor. The sensitivity of gate oxide structures to primary electron injection limits the use of EBIC on MOS ICs. These device altering effects for primary electron injection can be annealed at relatively low temperatures[23] (about 4 hours at 125 °C), but the annealing process can alter the defect being analyzed and may not be prudent during a failure analysis.

6.3.2 Resistive Contrast Imaging (RCI)

RCI utilizes a SEM to generate a relative resistance map between two nodes of a passivated integrated circuit [24, 25]. The technique obtains resistance information by using the integrated circuit an a complex current divider, amplifying and displaying the small (na) currents induced by exposure of an integrated circuit to the electron beam of the SEM. The relative resistance between the electron beam position on the circuit and the test nodes determines the direction and amplitude of charge flow. This current is amplified and used to make a resistance map of the conductors. Unlike electron beam induced current (EBIC) imaging, the primary electrons penetrate only to the buried conductors and not to silicon junctions. In fact, the EBIC signal, through charge multiplication in junction regions, can be several orders of magnitude greater than the RCI current. Therefore EBIC must be avoided so that the RCI data is not masked. Under these conditions the spatial resolution of the RCI image is roughly equal to the passivation thickness the electron beam is penetrating through. If EBIC is not a concern, higher beam energies can be used to improve the spatial resolution. A schematic of an RCI system is shown in Figure 6.18. Beam penetration for RCI must be sufficient to inject electrons into integrated circuit conductors even though the passivated surface is negatively charged. Only resistance changes relative to the test nodes are recorded by the RCI technique. Therefore, proper node selection is required and conductors without direct paths to output pins cannot be tested.

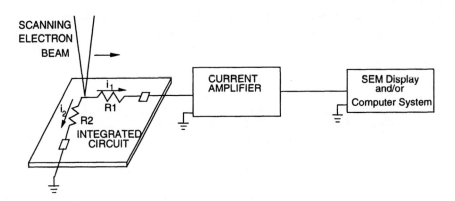

Figure 6.18. A schematic of and RCI system is shown.

RCI is primarily used to examine metal, polysilicon, and metal to silicon test structures and thin-film resistor conductor patterns. Figure 6.19 illustrates the use of RCI. The RCI image indicates an abrupt contrast change at a step with increased resistance. RCI requires no electrical stimulation for analysis, generates images that are quickly and easily interpreted, and permits examination of passivated and depassivated structures. RCI is very effective on simple test structures, but because only resistance changes relative to the nodes connected to the input of the RCI current amplifier are displayed. The use of RCI on ICs is limited and can be difficult to interpret.

6.3.3 Charge-Induced Voltage Alteration (CIVA)

CIVA[14] is analogous to electron beam induced current[12] (EBIC) and optical beam induced current (OBIC) imaging[11] in that the biased IC itself is the detector and amplifier. The signal monitored to produce a CIVA image is the power supply voltage with a constant current source used to bias the IC as an electron beam is scanned across the device surface.

To produce the CIVA signal, the primary electron beam must penetrate to the conductors of the IC as shown in Figure 6.20. If a conductor is electrically open, the charge injected into the conductor by the primary electron beam can reduce the voltage of the conductor and greatly affect the voltage demands of an IC supplied by a constant current source. CIVA uses these changes in a constant current power supply voltage with changes in gate bias to produce an image. The CIVA images display the conductors which are susceptible to voltage change by small amounts of injected charge, i.e. open conductors. The bias configuration of an IC for CIVA examination may be any non-contentive state such as that used in burn-in and life testing.

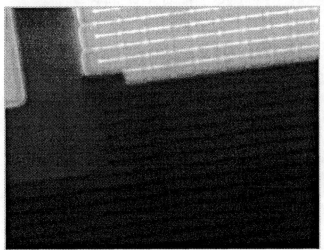

Figure 6.19. RCI image of a resistive contact (45 kΩ versus nominal 6 kΩ) metal-silicon serpentine test structure. To penetrate through the surface passivation a 9 keV primary electron beam energy was used.

While surface emission products like secondary electrons are not used to produce the CIVA image, they are a factor in image formation on passivated ICs. In order to penetrate through typical passivation layers the primary electron beam energy must be increased to 5 keV or above. At these beam energies more electrons are injected into the passivation layer than escape and a negative potential will build up on the surface[8]. This negative charge on the surface of CMOS ICs can effectively put all of the transistor gates into a low state, eliminating all IC functionality. As the primary electron beam energy is increased the interaction volume reaches buried conductors and a new current path is produced through the conductors. Charge can leave the passivation layer through the biased IC conductors and substrate. The additional current paths reduce the negative charge of the passivation and permit normal operation of the IC. Voltage contrast observation may still be obscured by surface charging from areas with passivation thicker than that over the conductors, but the CIVA signal is not affected by this charging. The small added amount of charge from the electron beam has little effect on the IC operating characteristics of non-failing conductors[26].

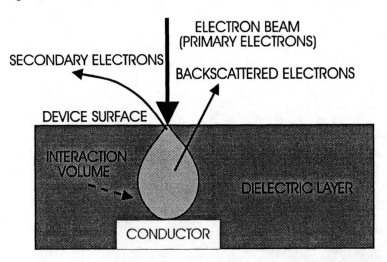

Figure 6.20. Interaction products of a 10 keV primary electron beam incident on a passivated IC.

To avoid radiation damage care must be taken not to use higher electron beam energies than are needed to reach the conductors of interest. The proper primary electron beam energy for CIVA imaging on passivated ICs is selected by first biasing the IC with a constant voltage source and the electron beam off. The IC current is then monitored as the primary electron beam energy is increased. The entire IC die is scanned at a rapid rate as the primary electron beam energy is increased. Initially the IC current will increase as the surface charges negatively. The IC current will decrease when the primary electron interaction volume intersects the buried conductors.

Once the proper primary electron beam energy is selected, the IC current[6] (I_{DDQ}) under constant voltage conditions is recorded. A current value slightly (5%) less than this is then used to operate the IC with the constant current source. The resulting voltage powering the IC is very nearly that used under constant voltage source conditions. A compliance voltage limit prevents any accidental damage to the IC from the constant current source.

As in LIVA, CIVA (and LECIVA) can take advantage of the "over-supply" biasing approach to improve the bandwidth response of the detection system. Using the "over-supply" approach TV scan rates are possible with magnifications showing an entire IC die for examination.

A block diagram of a typical CIVA system is shown in Figure 6.21. This is similar to the system used for LIVA and LECIVA.

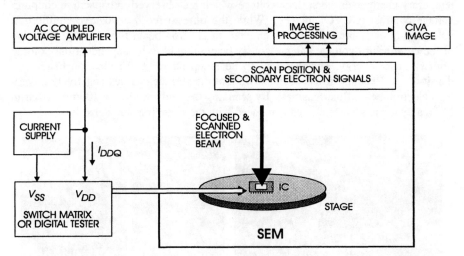

Figure 6.21. CIVA acquisition/processing system is shown.

Figure 6.22 illustrates the application of CIVA to a passivated CMOS gate array. The image was made using a primary electron beam energy of 15 keV. The high magnification image in Figure 6.22 shows that a CIVA signal is not produced for metal-1 when it is covered by metal-2. As with EBIC, CIVA can probe through optically opaque materials by increasing the primary beam energy. This "depth profiling" capability occurs because the penetration depth of the primary electron beam can be varied from approximately 0 to 6 μm by changing the beam energy from 0.3 to 30 keV respectively. Higher beam energies will also increase the probability of gate oxide damage from the primary electrons and should not be used unless needed for analysis.

6.3.4 Low Energy CIVA (LECIVA)

LECIVA[27] produces images of open interconnections through interlevel dielectric and passivation layers similar to CIVA images. The main difference between the two techniques is that LECIVA is performed at low primary electron beam energies ~ 1.0 keV.

The surface of passivation on an IC reaches a positive equilibrium voltage when the beam energy is between two cross over energies[28], E_1 and E_2. Typical values of E_1 and E_2 for insulators[29] are 100 eV and 3.0 keV. The positive surface voltage (normally ≤ 3 V) is due to more secondary and backscattered electrons being emitted from the dielectric surface than are incident from the primary beam. The charge imbalance generates a net positive voltage that retards the escape of lower energy secondary electrons until an equilibrium between incident and escaping charge is achieved. Changes in the voltage of conductors below the passivation produce temporary changes in the surface voltage which are observed in capacitive coupling voltage contrast (CCVC) images. While the time to reach a positive equilibrium voltage is directly proportional to the incident electron beam flux density, the value of the equilibrium voltage has previously been thought to be dependent only on the primary beam energy, and therefore to be independent of the electron beam flux density[28, 29]. Recently it has been shown that the surface charges negatively at very high electron beam flux densities. By scanning the high current (> 100 nA) electron beam a dynamic surface charge is induced with the changing flux densities.

Figure 6.22. Combined CIVA and secondary electron signals at two magnifications from a floating fan-out network in this two-level metal, 1.0-micron technology. A focused ion beam cut producing the open conductor can be seen in the right image.

For LECIVA the primary electron beam does not directly interact with the buried IC conductors. As illustrated in Figure 6.23, a changing surface equilibrium voltage will occur as the beam scans across the IC surface, dynamically polarizing the dielectric layer and producing a changing bound charge at the metal conductor-dielectric interface. The changing bound charge at the dielectric-conductor interface will introduce an ac voltage on the conductor. Similar to CIVA, the small ac voltage has little effect on non-defective conductors, but can change the voltage of open conductors, altering the power demands of the IC under test. Low primary beam energies ($\leq$ 1.0 keV) and high beam currents ($\geq$ 100 nA) should be used. LECIVA images are produced by displaying the changing V_{DD} of the constant current supply powering the IC as the beam is scanned over the IC. As with CCVC, the thicker the dielectric layer between the conductor and the surface, the smaller the bound charge effect will be.

For LECIVA signal generation, the changing bound charge on the conductor induced by the changing surface equilibrium voltages must alter the voltage of the electrically floating conductor. This implies that the dynamic charge from LECIVA must be greater than the charge loss by other mechanisms. One way to increase the frequency of the dynamic LECIVA charge is to use higher electron beam scan frequencies. Higher scan frequencies produce a larger LECIVA signal, but defect and amplifier bandwidth limitations may make the image appear smeared. Higher spatial resolution may be achieved by slowing the electron beam scan and operating the IC at a clock frequency faster than the scan rate. At slow scan rates the power demand of the clocked IC will be different when the electron beam is scanned above the tunneling open conductor. Thus, slow scan rates that produce better spatial resolution images can be used.

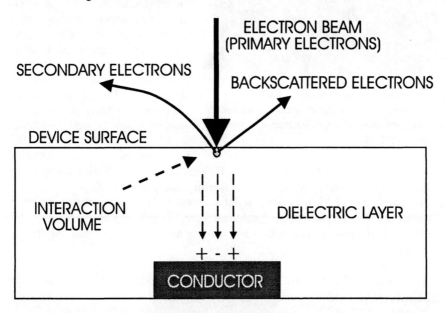

Figure 6.23. For LECIVA, the electron beam interaction volume does not reach the conductor. Polarization produces a changing bound charge at the conductor-dielectric interface.

For LECIVA signal generation, the changing bound charge on the conductor induced by the changing surface equilibrium voltages must alter the voltage of the electrically floating conductor. This implies that the dynamic charge from LECIVA must be greater than the charge loss by other mechanisms. One way to increase the frequency of the dynamic LECIVA charge is to use higher electron beam scan frequencies. Higher scan frequencies produce a larger LECIVA signal, but defect and amplifier bandwidth limitations may make the image appear smeared. Higher spatial resolution may be achieved by slowing the electron beam scan and operating the IC at a clock frequency faster than the scan rate. At slow scan rates the power demand of the clocked IC will be different when the electron beam is scanned above the tunneling open conductor. Thus, slow scan rates that produce better spatial resolution images can be used.

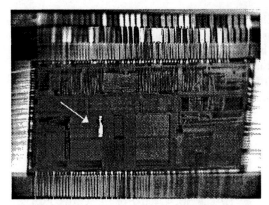

Figure 6.24. Combined LECIVA and secondary electron image of a microprocessor with an open conductor. The arrow indicates the open conductor.

While LECIVA has the advantage over CIVA of virtually no irradiation damage[22] and has the capability of being performed on commercial electron beam test systems, there are two major limitations. First, LECIVA cannot be used to detect open conductors which are completely covered by an upper metal layer. The upper conductor will block any charging of the lower level conductor. Second, as alluded to above, the LECIVA response is scan rate dependent. Commercial electron beam test systems usually scan only at a rapid TV rate producing a smeared LECIVA image across the field of view. Failure sites are localizable under these conditions but clear image recording of the floating conductor is difficult.

An example of LECIVA imaging on a three level metal microprocessor is shown in Figure 6.24. The primary beam conditions were 1.0 keV and approximately 150 nA. The open conductor in Figure 6.24 is a floating metal-2 segment resulting from a defective metal-1 to metal-2 via.

6.4 FUTURE DEVELOPMENTS FOR PHOTON AND ELECTRON BASED FAILURE ANALYSIS

PEM continues to be a powerful method for localizing IC defects. Careful analysis of the physics behind the different light emission processes from the most commonly encountered defect classes indicates there is a significant advantage in using the light emitted at wavelengths longer than 1000 nm compared to shorter wavelengths except for gate oxide defects. Detection of emission beyond the transmission edge of Si in the NIR will also be important as backside analysis continues to grow in importance due to flip-chip and multilevel metallization trends.

OBIRCH, OBIC, EBIC, and all of the "IVA" based techniques (LIVA, TIVA, SEI, CIVA, and LECIVA) are powerful techniques for IC failure analysis. For many of the photon based techniques an increase in defect sensitivity is desirable, especially on ICs with elevated power demands where the defect signal can be "lost" in the background "noise". Two improvements to increase the defect signal strength in active photon systems is increased photon intensity and pulsed photon sources. While a more powerful laser source will increase the signal, SOM optics limit the power that can be transmitted without damaging the optics. A pulsed laser and lock-in amplification of the active photon signals can also improve defect sensitivity. Although silicon has a high thermal conductivity, there are limits to how fast a laser pulse can be and still produce a useful change in temperature between pulses. Modeling indicates that the time constant for silicon surface heating by a focused laser is in the range of 100 to 200 ns. The thermal response of the surface to a 1 MHz modulated laser should produce changes in temperature within 90% of the steady state heating conditions. Therefore pulsed laser examination of active thermal deposition should increase defect sensitivity.

Challenges for these techniques include incorporation of the existing IC test fixture infrastructure, and CAD database. In addition to the equipment enhancements and automation, improving the detection sensitivity of CIVA and LECIVA for resistive, but not open, interconnections is required to address the challenges of shrinking contact dimensions. This is likely to require the incorporation of dynamic IC testing and possibly analysis of functional and timing parameters for fault localization and diagnosis.

Because of their great effectiveness in failure analysis and future potential, photon and electron probing, both active and passive, will be mainstays in IC failure analysis for the foreseeable future.

ACKNOWLEDGEMENTS

Sandia is a multiprogram laboratory operated by Sandia Corporation, a Lockheed Martin Company, for the United States Department of Energy under contract number DE-AC04-94AL85000.

REFERENCES

1   Hawkins CF, Soden JM, Cole Jr. EI, Snyder ES. The Use of Light Emission in Failure Analysis of CMOS Ics. Proceedings International Symposium for Testing and Failure Analysis, 1990, 55.

2   Soden JM, Cole, Jr. EI. Selected Topics in IC Failure Analysis: Light Emission Microscopy, Scanning Electron Microscopy Techniques, and Issues Concerning High-Pin Count Gate Array Devices. Tutorial Notes of International Reliability Physics Symposium, 1992, 4a1-4a.16.

3   Pankove JI. *Optical Processes in Semiconductors,* Ch. 6. New Jersey: Prentice-Hall, 1971.

4   Chynoweth AG, McKay KG. Photoemission from Avalanche Breakdown in Silicon. Phys. Rev., 1956, 102 (2), 369.

5   Soden JM, Treece RK, Taylor MR, Hawkins CF. CMOS IC Stuck-Open Fault Electrical Effects and Design Considerations. Proceedings International Test Conference, 1989, 423.

6   Henderson CL, Soden JM, Hawkins CF. The Behavior and Testing Implications of CMOS IC Logic Gate Open Circuits. Proceedings International Test Conference, 1991, 302.

7   Hawkins CF, Soden JM, Righter AW, Ferguson J. Defect Classes – An Overdue Paradigm for CMOS IC Testing. Proceedings International Test Conference, 1994, 413.

8   Kux A, Lugli P, Ostermeir R, Koch F, Deboy G. Infrared and Visible Emission from Si Layers.Proceedings Materials Research Society Symposium, 1992, 256, 223-226.

9   Shivanandan K, Nyunt K. Application of an Infrared Astronomy Camera System for Photoemission Microscopy. Proceedings International Symposium for Testing and Failure Analysis, 1995, 69.

10   Aw SE, Tan HS, Ong CK. Optical Absorption Measurements of Band-gap Shrinkage in Moderately and Heavily Doped Silicon. Journal Physics: Condens. Matter, 1991, 3, 8213.

11   Wills KS, Lewis T, Billus G, Hoang H. Optical beam induced current applications for failure analysis of VLSI devices. Proceedings International Symposium for Testing and Failure Analysis, 1990, 21.

Global Techniques

12 Cole Jr. EI, Bagnell Jr. CR, Davies BG, Neacsu AM, Oxford WV, Propst RH. Advanced Scanning Electron Microscopy Methods and Applications to Integrated Circuit Failure Analysis. Scanning Microscopy, 1988, 2, 133.

13 Cole Jr. EI, Soden JM, Rife JL, Barton DL, Henderson CL. Novel Failure Analysis Techniques Using Photon Probing With a Scanning Optical Microscope. Proceedings International Reliability Physics Symposium, 1994, 388.

14 Cole Jr. EI, Anderson RE. Rapid Localization of IC Open Conductors Using Charge-Induced Voltage Alteration (CIVA). Proceedings International Reliability Physics Symposium, 1992, 288.

15 Nikawa K, Inoue S. Various Contrasts Identifiable From the Backside of a Chip by 1.3 µm Laser Beam Scanning and Current Changing Imaging. Proceedings International Symposium for Testing and Failure Analysis, 1996, 387.

16 Nikawa K, Inoue S. New Capabilities of OBIRCH Method for Fault Localization and Defect Detection. Sixth Asian Test Symposium, 1997, 219.

17 Cole Jr. EI, Tangyunyong P, Barton DL. Backside Localization of Open and Shorted IC Interconnections. Proceedings International Reliability Physics Symposium, 1998, 129.

18 Ashcroft NW, Mermin ND. *Solid State Physics*. Ch.1, 24-25. Philadelphia: Saunders College, 1976.

19 Internet site, http://www.seebeck.com/seebeff.html, referenced 1/7/98.

20 Koyama T, Mashiko Y, Sekine M, Koyama H, Horie K. New Non-Bias Optical Beam Induced Current (NB-OBIC) Technique for Evaluation of Al Interconnects. Proceedings International Reliability Physics Symposium, 1995, 228.

21 Propst RH, DiBianca FA, Bagnell Jr. CR, Cole Jr. EI, Davies BG, Johnson DG, Lipkin LL, Neacsu AM, Oxford WV, Smith CS. Quarterly and Annual Reports to the SRC, Contract No. 83-01-025, 9/83-9/86.

22 Gorlich S, Kubalek E. Irradiation Effects on Passivated NMOS-Transistors Caused by Electron Beam Testing. Microelectronic Engineering, 1983, 1, 93.

23 Fleetwood DM, Winokur PS, Schwank JR. Using Laboratory X-Ray and Cobalt-60 Irradiations to Predict CMOS Device Response in Strategic and Space Environments. IEEE Transactions on Nuclear Science, 1988, 35 (6), 1497.

24 Cole Jr. EI. Resistive Contrast Imaging Applied to Multilevel Interconnection Failure Analysis. Sixth IEEE VLSI Multilevel Interconnection Conference, 1989, 176.

25 Smith CA, Bagnell Jr. CR, DiBianca FA, Cole Jr. EI, Oxford WV, Propst RH. Resistive Contrast Imaging: A New SEM Mode for Failure Analysis. IEEE Transactions on Electron Devices, 1986, ED-33 (2), 282.

26 Hawkins CF, Soden JM. Reliability and Electrical Properties of Gate Oxide Shorts in CMOS ICs. Proceedings International Test Conference, 1986, 443.

27 Cole Jr. EI, Soden JM, Dodd BA, Henderson CL. Low Electron Beam Energy CIVA Analysis of Passivated ICs. Proceedings International Symposium for Testing and Failure Analysis, 1994, 23.

28 Reimer L. *Image Formation in Low-Voltage Scanning Electron Microscopy*. Bellingham: SPIE Optical Engineering Press, 1993.

29 Kanaya K, Ono S. *Electron Beam Interactions with Solids, Scanning Electron Microscopy*. AMF O'Hare, 1984, p. 69.

7

PROBING TECHNOLOGY FOR IC DIAGNOSIS

Christopher G. Talbot, Ph.D.
Schlumberger

During integrated circuit diagnosis, contact and non-contact probing serves two key functions, namely fault localization and electrical characterization. As a fault localization capability, probing is usually the technology of last resort. Systematic back-tracing through a circuit (probing from a failing output back into the circuit to isolate the origin of the failure), while inherently capable of localizing almost any failure, is tedious and often quite time consuming. Fault localization techniques such as liquid crystal, emission microscopy, electron beam and optical beam induced current or voltage alteration techniques are often easier and quicker to use (See chapters 5 and 6). While these techniques are easy to apply they do not address all types of problems. For example, in design debug applications no physical defect usually exists in the circuit and back-tracing is frequently the only practical option. The back-tracing process is often facilitated by some level of simulation, test-based pre-localization software or image-based fault localization.

Electrical characterization is a key step in the failure analysis process. Once a failing component or net is identified, it is essential to understand in detail its electrical characteristics. This is commonly performed with mechanical probes and often includes some isolation, which allows characterization of individual failing components. Isolation is achieved by cutting metal interconnect lines using the mechanical probe itself or, more recently, laser ablation or FIB (Focus Ion Beam) technology.

7.1 PROBING APPLICATIONS AND KEY PROBING TECHNOLOGIES

Probing can be fairly well defined as the collection of electrical measurements from an IC's internal nodes. These measurements can broadly be grouped into three categories.

The simplest are DC measurements, for example, the measurement of the voltage on a single node. Voltage accuracy, either DC or AC, is frequently a vital element of circuit analysis particularly for analog devices, mixed signal circuits and memory sense amplifiers. When circuit loading is not a factor, mechanical probes offer a voltage accuracy advantage over non-contact technologies such as e-beam probing.

114   Probing Technology

AC probing is the second category and consists of both timing and waveform measurement. This type of measurement is primarily impacted by the bandwidth of the probing system. Since mechanical probes have significant loading characteristics (~100s fF), non-contact or loadless probing technology offers significant advantages, particularly for applications requiring very precise timing measurement such as design debug. Non-contact technology is also typically easier to use than mechanical probing but with limited voltage accuracy.

The third category is parametric measurement, which is typically performed as electrical characterization of a failing component, usually using two or more mechanical probes. Parametric measurements include a range of electrical characteristic from simple resistance and junction diode I-V (current versus voltage) curves to detailed transistor performance characteristics.

Measurement of internal circuit nodes can be performed in several ways. Mechanical and e-beam probing are the most common. Derivative techniques from Atomic Force Microscope have also been demonstrated, and are in a research and development phase. OBIC and EBIC, discussed in Chapter 6, are also capable of providing state information. Table 7.1 compares the advantages and limitations of the most common probing and fault localization technologies.

Technology	Risetime	Load	Primary Advantages	Key Limitations	Comments
Mech. Probe	~100ps	100fF	DC, Single shot, V, I, R	Damage, <~0.5um	Analog, FA tool
OBIC	NA	leakage	Fault Isolation	Access, no timing	Logic state only
EBIC	NA	NA	Fault Isolation	Damage, access	Logic state only
E-Beam	<50ps	0fF	Bandwidth, zero load	GFE, LFE, 10mV	Ideal for debug

Table 7.1. Top level comparison of probing and fault localization technologies is shown.

7.2 MECHANICAL PROBING

Mechanical probing is the lowest cost, most versatile, and oldest probing technology. It has been in use since the days of the earliest integrated circuits. Its most significant advantage is direct electrical contact, which can be used to accurately measure voltage, current and impedance. Direct contact probing also allows non-sampled, single shot AC measurements up to the bandwidth limit of the probe. This is important as single shot AC measurements cannot be made using non-contact, sampling-based probing technologies.

Use of multiple mechanical probes allows simultaneous measurements at several locations in parallel, another capability that is not possible with non-contact technology. It also allows signal forcing or injection while making a measurement. This is essential in electrical characterization since the failed component must frequently be electrically isolated from the rest of the circuit for unambiguous and accurate characterization.

Mechanical probing is intuitively simple. Mechanical probing using an optical microscope does not require an expensive vacuum system with all the associated fixturing and heat sink complications. A wide variety of commercial probe tips are available for a broad range of applications.

7.2.1 Mechanical Probing Technology

A typical mechanical probe station consists of five key elements: a microscope, stage, probe manipulator, probe tip and device stimulus. Historically, each of these elements was quite simple. As integrated circuit technology has advanced, each of these elements has become more complex.

The microscope must have sufficient resolution to image the conductors to be probed and enough depth of field to follow the probe tip movement. For optical systems, different objective lenses are usually provided to accommodate different magnifications and depths of field. Lower magnification with greater depth of field is essential for coarse location of the probe tip. Advanced sub-micron IC's, however, require higher resolution microscopes in order to image the conductors. This drives the use of higher resolution microscopes such as SEM and FIB for probing. A manual positioning stage is sufficient for basic mechanical probing in simple setups. However, a precision motorized XY stage with accuracy on the order of a few microns is typically required for VLSI devices. The precision mechanical stage is frequently linked to a workstation to allow the use of CAD navigation software to reduce the time required to locate the signals to be probed. Mechanical stage accuracy and repeatability are particularly critical for repetitive structures such as memory. If a stage cannot arrive at a designated memory bit with certainty, the tedious and time consuming process of cell counting must be employed.

For simpler applications, manually controlled micro-positioners are sufficient for positioning the probe tip on older IC technologies. VLSI probe systems employ precision motorized positioners under computer control including software linked to a sensor to aid in the tip touchdown. Piezo vernier motors provide extremely fine positioning resolution (~10nm) for probing submicron features.

Etched tungsten wire is the probe tip material of choice. It is hard which is critical to tip lifetime. It can be etched to provide tip radii as small as 0.1-0.2um for probing small features. The spring constant of advanced probe tips is usually designed to limit contact pressure to minimize the chance of device damage. For AC probing a FET amplifier is integrated close to the end of the tip to minimize loading and thereby increase substantially measurement bandwidth. AC probes offer giga-hertz bandwidths with low effective loading capacitance down to ~100fF and high load resistance, usually >1MΩ. For RF applications, even higher bandwidths can be achieved at the expense of higher loading capacitance and lower load resistance.

Mechanical probe stations can usually use the existing stimulus setup or tester load board to drive the device being probed (see Chapter 2). Both packaged dice and wafers can be probed.

116   Probing Technology

7.2.2 Mechanical Probing Limitations

As indicated above, mechanical probing provides unique capability for probing DC and one-shot events as well as for characterization probing. The effective use of mechanical probing requires an understanding of its limitations, which are significant.

Mechanical damage from time to time is inevitable when making mechanical contact directly with the microscopic features on the active face of an IC. This is increasingly a problem as design rules shrink below one micron. Beyond using carefully designed probe tips, and having the analyst take great care when probing, there is little that can be done to eliminate this risk.

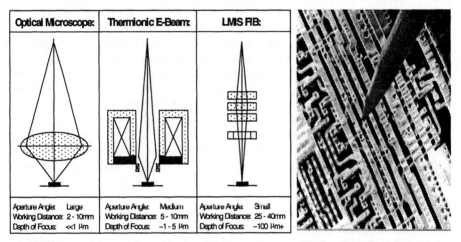

Figure 7.1 Depth of focus comparison diagram is show with an example FIB image showing a mechanical probe and large depth of focus.

As metal interconnect design rules move into the deep sub-micron regime, both the optical diffraction limit of resolution and small depth of focus (particularly with high resolution high numerical aperture lenses) of optical microscopes make it extremely difficult to see or probe deep submicron lines with conventional mechanical probe systems. Probe-points, either designed-in during the development process or deposited after the fact using a FIB system, are the most common solutions to this problem. Other than the impact on diagnostic productivity, FIB probe-points essentially allow electrical contact with all conductors that the FIB system can access. Use of e-beam or ion beam microscopes overcome both the resolution and depth of focus issues (see Figure 7.1) but adds significantly to the cost of mechanical probing. Nevertheless the combination of a FIB (Focused Ion Beam) system for in-situ probe-point creation and repair with an integrated mechanical probe capability allows direct probing of deep submicron features relatively easily and with substantially reduced risk of damage[1].

Since most metals oxidize when exposed to air and moisture, the conductors inevitably are covered with a thin layer of native oxide that must be scraped away or broken through with the aid of the probe before reliable ohmic contact can be made. Incomplete removal of passivation or dielectric can make this problem worse. Experts in mechanical probing often attempt to push the probe tip into the side of conductors in order to reliably make electrical contact while minimizing the chance of damage due to the probe sliding. In the best case, a residual contact resistance of a few ohms is typical, simply due to the relatively small contact area. A few ohms of contact resistance is of no consequence for applications where the current through the probe is minimal, as is the case for most voltage measurements. However, for low resistance measurements in particular, contact resistance can limit accuracy unless four point probe techniques are employed as shown in Figure 7.2.

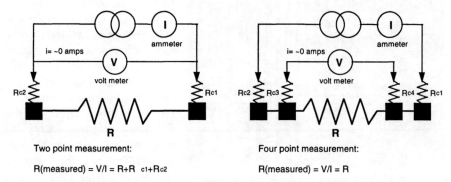

Figure 7.2 Four point probing eliminates errors from contact resistance. The voltage across the resistance of interest is measured directly with separate probes in a separate zero current circuit. As virtually no current flows in this separate circuit, errors due to voltage drop of across the contact resistance are eliminated.

Electrical loading (capacitive and resistive) is primarily an issue for high bandwidth timing measurements especially for sensitive or weakly driven signals such as charge storage nodes or dynamic logic. The actual magnitude of the effect depends of probe size and its electrical properties relative to those of the node being probed. Even with the high performance AC mechanical probe tips, there is frequently sufficient uncertainty in the exact error introduced by loading that a non-contact, non-loading probing technology, such as e-beam, must be employed.

Even given these limitations, mechanical probing is clearly a technology that is here to stay because of its inherent low cost and extreme flexibility for a very wide range of applications.

7.3 E-BEAM PROBING

E-beam probing uses the natural phenomenon of voltage contrast in a Scanning Electron Microscope (see chapter 10). One of the first demonstrations[2] of this was in

1956. A conventional SEM image is produced by rastering (raster scanning) a finely focused beam of primary electrons over a specimen. Secondary electrons, with relatively low energies 0-~15eV, are produced and detected. The resulting signal is used to form an image of the specimen. Voltage contrast is a phenomenon, which produces an intensity variation in the SEM image that is a function of the surface electrical potential as shown in Figure 7.3. The emission of secondary electrons is modulated by the local electric fields on the DUT (Device Under Test). A positive voltage (shown in the right half of Figure 7.4) on the surface of the DUT reduces the secondary electron count while a zero or negative voltage does not (shown in the left of Figure 7.4). Thus, positive voltages appear as darker areas in the image, corresponding to fewer secondary electrons emitted, while zero or negative voltages appear as light areas and correspond to higher secondary electron emission[3]. The modulation of the secondary electron signal is a relatively strong effect, as the secondary electron energy or potential is comparable in magnitude to the surface voltage on a typical IC.

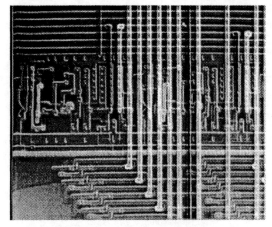

Figure 7.3 A voltage contrast image shows surface DC and buried AC voltage contrast.

Raw, real-time voltage contrast images themselves can be very powerful diagnostic tools. For example, a simple fault in a chip such as a broken line stands out as clearly as night from day as the voltage contrast across the open circuit clearly identifies the failure location.

SEM's optimized for e-beam probing are equipped with high-speed beam pulsing hardware (often referred to as a beam blanker). They are capable of pulsing or strobing the beam with pulse widths from 50ps to several microseconds. This is critical for high speed signal acquisition and greatly enhances the measurement bandwidth as discussed below.

Directing a pulsed e-beam at a particular node of interest provides a mode of operation very much like that of a sampling style oscilloscope and enables the acquisition of high-speed quantitative voltage waveforms with an equivalent bandwidth exceeding 10GHz. Timing information and repetitive dynamic faults are readily characterized.

E-beam probing has several advantages in addition to the basic modes of voltage contrast imaging and waveform acquisition. Specifically, high spatial resolution (typically <0.1um), non-contact, non-loading, high bandwidth, non-destructive (providing the beam energy is <1.5keV) and the ability to probe buried conductors.

Buried conductor probing is possible because the material covering the conductor acts as a dielectric and capacitively couples the potential at the DUT surface to that of the buried conductor. As the e-beam acts as a loadless probe, it is capable of measuring this capacitively coupled signal. The actual equivalent circuit of the e-beam is an ideal or infinite impedance current source that injects an average current of between one thousandth and one millionth of the primary beam current of 1-10nA.

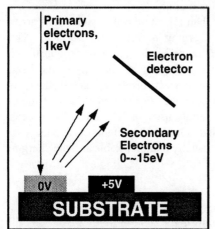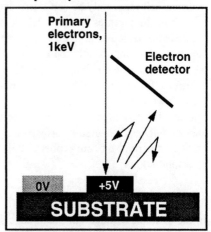

Figure 7.4 The voltage contrast phenomenon principles are demonstrated.

7.3.1 Surface Charging and Beam Energy

Optimization of the e-beam conditions for probing is a critical part of making efficient and accurate measurements. Stable surface charging is the first priority. In general, achieving stable surface charging in a SEM is not fully understood and is something of an art rather than a science. The secondary electron yield curve (Figure 10.9) is a characteristic of the specimen material, particularly the surface condition or surface work function. For most materials the general shape is as shown.

Two so-called crossover energies (also called crossover voltages), often referred to as E1 and E2, are shown where the yield curves crosses the unity line, indicating that the primary and emitted secondary electron currents are equal. In principle, in this condition every primary electron delivered to the surface is balanced by one electron leaving the surface and (per Kirchoff's current law) the surface potential will remain constant as no net charge is being deposited or removed. In practice, the exact crossover energies will vary slightly depending upon the exact surface condition, which can result in unstable charging even close to these crossover energies.

If the beam energy is greater than E1 but less than E2, the secondary electron yield is greater than one. This means that more electrons are leaving the surface than are arriving. Net removal of negatively charged electrons is equivalent to delivering positive charge to the surface and results in an increase in the surface potential. This increase will in turn produce an electrostatic retarding field at the surface of the device that reduces the number of secondary electrons escaping (assuming that the SEM or prober is not perturbing or swamping this field with a strong electrostatic collection field). Thus, a stable dynamic equilibrium condition is readily achieved, as the surface will charge positively to a couple of volts until the secondary current exactly balances the primary beam current. Any slight variations in surface work function will inherently be self-correcting in this regime.

For beam energies less than E1 or greater than E2, more primary electrons arrive than electrons leave and thus net negative charging occurs at the surface. This negative charging is inherently unstable and in principle, for a perfect insulator, would eventually result in the surface potential charging all the way to the beam energy. In practice ICs are not made with perfect insulators and leakage or breakdown will occur, potentially damaging the device. For silicon oxide and silicon nitride, the two most common dielectric materials used in ICs today, 1keV-beam energy will result in stable surface charging and is hence the default beam energy for most commercial e-beam probers. Polyimide usually exhibits stable surface charging at around 800eV.

7.3.2 Buried Conductors, Capacitive Coupling and Limitations

In the example voltage contrast image shown in figure 7.3, you can clearly "see" not only surface metal-2 conductors, with DC voltage contrast, but also buried metal-1 conductors. The vast majority of secondary electrons escape from the top few tens of nanometers of the surface and are influenced by the surface only. The buried conductors that we "see" influence that surface in two ways, namely through induced surface topography, and because of capacitively coupled AC voltage contrast resulting from a signal present on the buried conductor.

The induced surface topography gives rise to an increased angle of incidence for the primary beam which results in preferential generation of secondary electrons and a bright edge in the image. The human eye, which is tuned to enhance edges and fill-in in between, sees these bright edges and "sees" a buried feature. Note that this preferential secondary electron generation phenomenon can result in substantial voltage amplitude errors if the probe beam is placed too close to a significant topographical edge during waveform acquisition.

Capacitively coupled voltage contrast is very valuable. Under stable surface charge conditions, it allows excellent AC waveform acquisition from buried conductors without the need for dielectric removal or probe-point creation. Note that DC signals cannot be acquired from buried features just as a DC measurement cannot be made through a capacitor. Although the actual capacitance from the buried conductor to the surface is minute, e-beam probers are able to make this type of AC measurement effectively as they are virtually ideal, loadless probes.

Historically, probing buried conductors by capacitive coupling was commonplace. With the increasing number of layers of interconnect and increasing conductor cross-sectional aspect ratio, capacitive coupling has become less useful for more advanced IC's.

7.3.3 Linearization For Quantitative Voltage Measurement

While a voltage contrast image does an excellent job of showing qualitatively surface potentials, the actual secondary electron signal is not a linear function of surface voltage. To measure quantitative voltage waveforms, this characteristic is linearized with a feedback or servo loop[4].

Figure 7.6 shows a typical voltage linearization system based on the servo-loop feedback principle. In this implementation, secondary electrons leaving the surface of the device are collimated by the magnetic field projected forward by the objective lens. Secondary electrons, regardless of their energy, will be collimated parallel to the optical axis provided the device is approximately at the point of maximum axial magnetic field[5]. Thus, the filter mesh or grid will act as a retarding field spectrometer and can discriminate electrons on the basis of their total energy and not just the uncollimated vector component of velocity perpendicular to the grid.

Figure 7.5. Five layer metal process where capacitive coupling will not work effectively for waveform measurement.

The number of low energy secondary electrons leaving the surface is highly dependent on surface potential (the basis of voltage contrast images). These low energy electrons are also the most susceptible to local field interference (known as Local Field Effect, LFE) which can result in crosstalk from neighboring signals. Crosstalk is minimized or eliminated by using the filter to reject lower energy electrons and using the higher energy electrons alone to determine voltage. The feedback loop adjusts the filter mesh potential to maintain the photomultiplier output current constant and equal to a fixed reference current. Under this condition, the filter electrode voltage will track the surface potential of the device. The waveform can be digitized and displayed accordingly, although there is a DC offset voltage that rejects lower energy secondary electrons.

Surface AC voltage measurements are typically accurate to better than 5% of their amplitude. DC measurements, which of course are only possible on exposed surface conductors, are inherently less accurate as a relative measurement must be made using a local reference point. In fact all DC measurements, contact or non-contact require a reference. Subtle variations in work function thus add an additional uncertainty to any DC measurement.

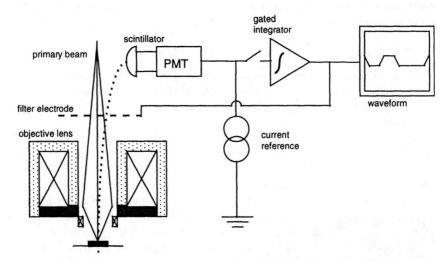

Figure 7.6 A typical voltage linearization system is shown.

Note that the integrator shown in Figure 7.6 is gated. The gate is switched off and the integrator acts as a hold circuit while the beam is blanked thus rejecting background noise and minimizing subsequent signal averaging. The integrator will have a time constant on the order of one to five microseconds and must accumulate signal from approximately that much cumulative beam "on" time to allow the control loop to settle and for the output voltage to be valid. A complete voltage waveform is acquired by sequentially incrementing the relative phase of the sampling pulse after the feedback loop has settled for each point in the waveform. More advanced systems use a dual gated integrator to further reduce noise by subtracting background noise and dark current from the integrator[6]. The integrator can of course be implemented in analog or digital form.

7.3.4 Sampling, Duty Cycle and Acquisition Speed

Figure 7.7 shows a typical timing diagram of the relationship between the test pattern and beam sampling pulses. If an integrator time constant of 1us is being used with a 1ns beam pulse width, several thousand beam pulses are required for the control loop to settle (typical for a commercial prober operating with 1-3nA of beam current). If we assume a test pattern repetition period of 10us and 1000 beam pulses per point in a 500 point waveform display, then 500,000 beam pulses are required and a single waveform sweep will take approximately 5 seconds, assuming minimal system overhead.

This simple example clearly illustrates the duty cycle limitation inherent in e-beam probing. Since everything scales linearly, a 100ps sampling pulse used in conjunction with a 100s test pattern will result in a ~500s or ~8 minute acquisition time (for a given beam current and target signal to noise ratio).

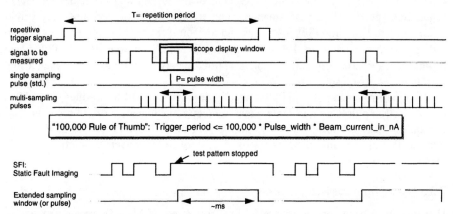

Figure 7.7 Timing diagram shows the relationship between test pattern and e-beam sampling pulses.

The simplest approach to improving the duty cycle limitation is to reduce the test pattern length. In some situations reproducing a particular failure mode will be straightforward and a small subsection of the test pattern can readily be used. However, in many situations reducing the test pattern length can be impractical.

Using multiple sampling pulses per test pattern execution, as illustrated in Figure 7.7, would seem to be a means of alleviating this duty cycle limitation. The maximum rate of multisampling is, however, limited by the bandwidth of the secondary electron detector, typically to one pulse every ~100ns. Multisampling accelerates acquisition speed for logic waveforms with longer beam pulse widths and wider acquisition windows. However, multisampling has no impact on the real practical acquisition time required for data acquired at the shortest sampling pulse widths as only a single sampling pulse can be placed in the short waveform time window due to the finite detector bandwidth and secondary electron flight time.

Ultimately, acquisition speed is limited by shot noise in the secondary electron signal. Shot noise is the statistical noise that results from the fact that the signal is not a continuous stream of charge but is quantified literally by the number of secondary electrons collected. The signal-to-noise ratio is proportional to the standard deviation of the number of electrons collected per point in the waveform and is given by the square root of the number of electrons. A 1ns primary beam pulse of a 1nA beam contains approximately six electrons. This will result, on average, in the generation of six secondary electrons, most of which are collected and used for imaging. For waveform acquisition, however, approximately half of secondary electrons have sufficient energy to be passed by the spectrometer and extensive signal processing and averaging is required to recover the signal.

Increasing beam current therefore allows faster acquisition. The actual current into a particular spot size is a function of the electron column design and the electron source. The "brighter" the source, the more beam current and the faster the signal acquisition speed. High performance commercial e-beam probers use very bright Thermal Field Emission (TFE) sources and are routinely able to deliver ~10nA in to a 0.1µm spot compared to 1-3nA for thermionic emission sources. A TFE beam results in waveform acquisition that is approximately tens times faster than using thermionic emission sources.

High performance pulsed UV photocathodes[7] can be used to produce very short, bright electron pulses with instantaneous beam currents in excess of 100nA with 10ps durations. This technology employs high power mode-locked lasers to generate the intense UV pulses required. This approach has not so far proven to be commercially viable.

7.3.5 Device Stimulus Requirements

Historically, e-beam probers were in fact converted SEMs. Fixturing the device inside the vacuum chamber and providing high bandwidth stimuli was time consuming and difficult. Current systems are optimized for e-beam probing. This optimization includes not only the electron beam and electron detector but also the configuration for easier interface to electrical stimulus. This is achieved by using the inverted moving column architecture (figure 7.8). In a conventional SEM, the stage is used to move the sample and the column remains in a fixed position. By moving the column rather than the device, electrical connections to the device remain static while scanning various areas of the DUT. The column is inverted to provide easy access to the vacuum chamber and in particular to the device pins. A device socket is mounted on a flange with the DUT inside the vacuum chamber and the pins outside. The device pins are connected directly through the module which provides a short, usually impedance controlled, connection to the electrical

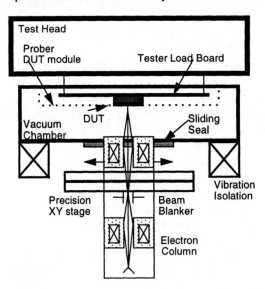

Figure 7.8 E-beam prober column and chamber section showing the inverted moving column architecture that allows easy access to the DUT for electrical and thermal stimulus.

stimulus. This makes possible direct docking to full function commercial ATE. This flange mounting approach also facilitates temperature control, which is critical for high power devices.

Wafer probing is accommodated by placing a probecard on the floor of the vacuum chamber and using a lower precision wafer stage in place of the DUT module to position a specific die on the wafer to be probed on the probecard.

7.3.6 Advantages and Limitations

E-beam probers have a wide range of advantages over mechanical probers: non-contact and therefore non-destructive, non-loading, high bandwidth, and easy to use. However, there are some important limitations.

The primarily limitation is that the e-beam prober functions as a sampling oscilloscope. In mechanical probing, any electrical measurement device (or stimulus for forcing) can be connected. E-beam probing suffers the same limitations as sampling oscilloscopes as well as some additional restrictions. Trigger synchronization is required for all sampling oscilloscopes. However, unlike a conventional sampling-style oscilloscope, an e-beam prober cannot trigger from the signal being probed. In most cases the tester or other stimulus source must provide a hard wired trigger signal. A reasonable trigger frequency for good speed is defined by the 100,000 rule-of-thumb (100,000*beam_current in_nA * pulse_width = max_trigger_period). The signal to be measured must have a fixed phase with respect to the trigger. Internal free running oscillators cannot be probed without a hardwired trigger signal. Intermittent signals cannot be measured, as the basic assumption required for equivalent time sampling is that the waveform to be measured is repetitive. Intermittent failures must first be forced to predictably repeat prior to using an e-beam prober. This can often be done simply by varying the temperature or power supply voltage. Stimulus source jitter ultimately limits the risetime or effective bandwidth of any measurements. As e-beam prober performance improves, stimulus source jitter stability, rather than fundamental measurement bandwidth is increasingly becoming the limiting factor in the equivalent measurement bandwidth.

The accuracy of voltage measurements is also limited. AC accuracy is limited by spectrometer linearity and signal averaging to ~5% of the AC amplitude of signal being measured down to ~10mV. DC accuracy is limited by local variations in surface work function (in addition to the AC limitations listed). All DC measurements are relative measurements, and subtle differences in surface conditions result in DC errors that cannot be corrected.

Other limitations in e-beam probing due to field effects on the device become more critical as geometries shrink. The primary e-beam can be deflected by changes in global electric fields and magnetic fields. These global fields are caused by bond wires, bond pads and global signal routing on the IC, and are becoming an increasingly significant issue. More advanced e-beam probers include software that predictively corrects global field effect (GFE) deflection by deflecting the beam prior to landing in order to correct the errors. Changes in local electric field experienced by secondary electrons, as they leave the device surface, can result in crosstalk and amplitude errors. The best solution to this problem is a more accurate secondary electron spectrometer (less angular dependence in the columnation process), since in principle, local field effects (LFE) cannot affect the total energy of the secondary electron - only their trajectories.

Smaller feature sizes also reduce the capability to make capacitively coupled measurements. Capacitive coupling is limited to the probing of conductors that are within one conductor width of the dielectric surface. Capacitively coupled crosstalk is a property of the device and can only be measured, not eliminated, by the probing technology. Use of probe-points is the best solution (either designed in or created with a FIB).

The integrity of the electrical stimulus provided for e-beam probing is critical. Decoupling capacitors should always be used on the power supplies (typically $0.1-0.5\mu F$ depending on the peak transient currents). A DUT board mounted ballast capacitor is also appropriate ($\sim 10\mu F$). All conductors on the DUT should be biased or grounded, since the electron beam can cause floating nodes to charge. Pull-down or pull-up resistors should be used on outputs that have the potential to float, e.g. tri-state, bi-directional, open collector, open emitter, open drain, etc. All conductors inside the chamber near the DUT should be tied preferably to ground or at least to one of the power supplies. AC signals must have fast, clean edges if accurate rise-times are to be measured.

7.3.7 Stroboscopic Imaging and Fault Imaging

So far, only real-time voltage contrast images and stroboscopic waveforms have been discussed. Stroboscopic images can also be collected by pulsing the e-beam at a particular point in the test pattern, incrementing the corresponding image pixel acquired (or historically using a SEM's slow-scan mode) and averaging the collected signal over time in a digital frame grabber. The result is an image that shows a snapshot of the state of the IC at a particular point in the test pattern. A strobe image of an area being probed can be a useful guide for waveform based diagnosis. Collecting strobe images from a sequence of consecutive points in the test pattern allows a "mini" movie of the operation of the IC to be produced.

Historically, stroboscopic imaging was relatively simple to implement on the persistent green phosphor SEM screens. These screens provided sufficient signal averaging and storage capability to show stroboscopic images at relatively short duty cycles. Simply incrementing the phase of the beam blanking pulse (Figure 7.7) allowed a crude but effective means of performing logic analysis and stepping through a circuit's operation, logic state by logic state.

Synchronizing the Y image scan signal with the phase or time increment of the beam pulse provides another historically useful imaging mode on a conventional SEM, referred to as a Logic State Map (LSM). Aligning the imaging area with a bus structure, for example, allows simultaneous acquisition of the state of a whole bus over a given period of time. As is the case with any stroboscopic imaging mode, acquisition speed is slow for longer test patterns. Stroboscopic Imaging has largely been supplanted by Fault Imaging Approaches discussed below.

Dynamic Fault Imaging, DFI, and Static Fault Imaging, SFI, are image-based e-beam fault localization techniques that use a fully functional reference device to allow diagnosis without detailed knowledge of device operation.

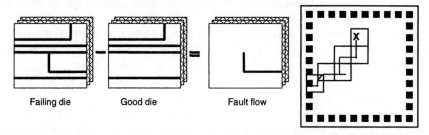

Figure 7.9 Fault localization with image differencing is illustrated.

DFI[8], as illustrated in Figure 7.9, refers to the technique of capturing a sequence of strobe images from a reference or golden device and the same sequence from the failing device. The two image sequences are then precisely aligned and differenced. The resulting difference image sequence shows the propagation of the fault backwards in time from the output pin to its origin.

For simple circuits with relatively short test patterns, fast and efficient fault localization is readily achieved with DFI. However, for large complex circuits with long test patterns, stroboscopic image acquisition becomes unacceptably slow. Complex internal bus structures can dramatically increase the number of difference images required as the fault is effectively fanned out across the chip by the bus. Advanced multi-layer metal ICs with large areas covered by power planes can eliminate virtually all failing node information from the surface without probe-points. As a result, DFI becomes more time consuming and impractical as devices grow in complexity.

SFI[9] addresses the strobe image acquisition speed problem by stopping the test pattern at the vector of interest and then extending the e-beam pulse to several milliseconds as shown in Figure 7.7. Image acquisition time is reduced to approximately 1-3 sec. per image and is more or less independent of test pattern length. As the name SFI implies, this technique does not allow characterization of dynamic faults. Backtracing of dynamic faults is possible, however, since most dynamic faults result in the propagation of static logic state errors. SFI relies on the device maintaining its state for several 10s of milliseconds after the test pattern has stopped and on the charge induced on the surface of dielectric by buried features not leaking away for this same period. These are both usually valid assumptions even for circuits that employ dynamic logic or other non-static charge-based logic.

128   Probing Technology

SFI eliminates the acquisition speed limitation of conventional strobe imaging. In practice, implementation of image sequence differencing becomes a matter of acquiring, aligning and comparing sequences of good and failing images. Another obstacle for DFI and SFI is that images must be acquired on a good device as well as the failing device. This can require many time consuming sample changes. In some cases, the failing device can be operated properly at another temperature or power supply voltage. Mechanical systems, which allow rapid interchange of good and failing devices, have also been constructed[9].

No attempt is made to solve the precision image alignment problem, but instead the two frame buffers are toggled (or flickered) back and forth. In spite of the lateral shift of features from one image to the other due to residual stage and alignment errors, the human eye has no problem discerning image differences. Back tracing to the origin of the fault is then simply a matter of "panning" the device viewing area, as the test vector being imaged is decremented one step at a time. The cycle for each image pair takes less than a minute and a fault on a medium complexity circuit that is not covered by power planes can usually be localized in a matter of 10-15 minutes.

7.4 NAVIGATION AND STAGE REQUIREMENTS

Historically, paper printouts of circuit layouts were used as maps for failure analysis. For ULSI devices, this is not practical and an automated approach for accessing these maps must be integrated into probing tools as well as other failure analysis tools. Navigation on these devices without design layout support is impractical.

Figure 7.10 shows an e-beam prober user interface that includes integrated CAD navigation software. The particular implementation shown includes both the physical CAD layout of the IC and its logical circuit schematic. Net list navigation is also critical. The major part of the power of these tools is their ability to tie the various representations of the device together with a physical image. In this example, the user simply clicks on a signal or net in the schematic or netlist display. The net name is identified by the software and is used as an index to a large lookup table that links net names and the corresponding polygons in the layout database. The layout display software uses

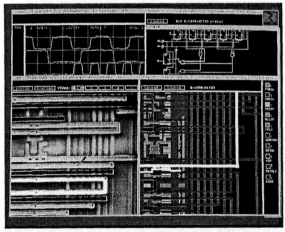

Figure 7.10 E-beam prober user interface with full CAD navigation is shown (Courtesy Schlumberger ATE).

this polygon information to highlight the list of corresponding polygons in the layout display and moves the display in X and Y to show the physical signal layout. The same X and Y move commands are also sent to the stage controllers and the physical XY stage (which requires pre-alignment to the CAD database) moves either the IC or the viewing optics to show the corresponding physical location on the IC. The result is rapid, accurate and virtually error-free navigation around the IC.

There are two primary pre-requisites to use CAD navigation, a pre-processed CAD database(s) and a precision XY stage[10]. The CAD navigation software provides a lookup table that links the physical layout, logical or schematic domains and physical analysis tool to reduce time-consuming real-time searches of the CAD database. Internally the layout database consists of a long list of polygons, each with its corresponding shape, size and XY location within the database and the layer it represents (diffusion, polysilicon, metal-1, metal-2 etc.) but without information about the electrical connectivity that they represent.

The netlist database is simply a textual list of hierarchical circuit elements, from simple transistors at the lowest level of the hierarchy to functional blocks at the top of the hierarchy, with a corresponding list of interconnections. A schematic database is essentially the same, except it also describes the graphical circuit elements that make it easier for humans to interpret the functionality depicted. As many designs today are implemented at the schematic or logic level using logic synthesis tools, graphical schematics are not available and a less convenient netlist display must be used for navigation. Some navigation software packages now have the ability to synthesize a graphical schematic from the netlist using a library of standard logic symbols and a simple place and route algorithm to display the wiring.

Both databases are typically required in an industry standard format such as GDS II or EDIF. Pre-processing consists of interpretation of the connectivity represented in the layout database and extraction of the effective netlist that it represents, mapping of the extracted layout netlist with the actual netlist used for simulation of the design and construction of the lookup tables for the run-time software. This process is very similar to design verification and in fact often makes use of commercial design verification tools.

In the general case of random logic ICs, stage requirements for navigation are not particularly demanding. The stage must be able to place the feature of interest in the field of view (FOV) of the microscope at a magnification, which enables the operator to recognize the feature selected. Typically, a stage accuracy of a few microns and a minimum step resolution of a few tenths of a micron is perfectly adequate.

For repetitive memory cells, stage and alignment accuracy is essential if an individual memory cell is to be uniquely identified. If the repeat distance of the memory cells is P as shown in Figure 7.11, then the overall microscope or beam positioning accuracy (including alignment errors) must be less than P/2, preferably ~P/4. High performance stages with interferometric position feedback are capable of positioning accuracy of a few tenths of a micron. It is, however, important to remember that total beam position error, not just stage position alone, is the real requirement. In some cases, large calibrated beam deflections referenced to a local alignment can be used to emulate or replace an interferometric stage[11] at a substantially lower cost. In some environments such as the e-beam prober, the stimulus source can be used to toggle the memory cell of interest. The voltage contrast image will then clearly indicate which memory cell is being addressed.

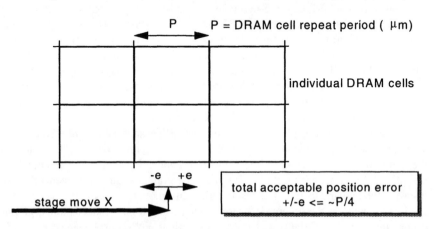

Figure 7.11 The positioning accuracy needed to identify directly an individual memory cell in an array is assessed.

7.41 CMP, Planarization and Overlay Techniques

Planarization eliminates surface topography and thus makes buried conductors effectively invisible to charged particle beam images. This is a problem for FIB images, as a frequent requirement is to drill accurately down to a lower level conductor between two upper level conductors without touching or damaging them.

This problem is solved by superimposing or "overlaying" an outline form of the layout display on top of the charged particle beam image, as shown Figure 7.12. The overlay is locally

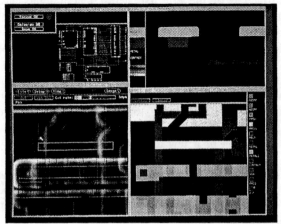

Figure 7.12. CAD layout overlay accurately identifies the location of invisible buried conductors.

aligned using visible surface topography and thus invisible buried features are located accurately or at least to within the tolerances of IC manufacturing.

When using an Infra Red (IR) optical image to perform diagnosis through the thinned reverse side of an IC, navigation and overlay are particularly important due to limited spatial resolution of IR microscopes. From an implementation viewpoint, a layout display tool capable of displaying a flipped or reversed layout is the primary requirement. Overlay with an IR optical image is also an important navigation technique for flip chip analysis.

7.5 FIB FOR PROBING AND PROTOTYPE REPAIR

Probe point generation using a Focused Ion Beam (FIB) has become a significant part of both mechanical and e-beam probing. The properties that make FIB ideal for signal access through probe-point creation also make it the ideal prototype repair technology. In one sense a repair is simply a connection between two or more probe-points usually combined with some signal line isolations or cuts. All of these operations require the same capabilities as probe-point creation. Such electrical modifications are themselves often part of the probing process as isolating a circuit or using a strap to connect a key missing signal are often essential for diagnosis. FIB technology has also found additional applications in cross section preparation and defect analysis.

Small feature sizes, multi-level metallization, and top level coverage with wide buses contribute to the difficulty in signal access and the need for probe-point generation. The number of signals, which need to be accessed, can be reduced by improved testability using techniques such as scan test and boundary scan[12]. These design methodologies increase both the controllability and observability of internal nodes through the external pins and can be used to pre-localize a failure with appropriate software prior to probing. The ability to probe signals and create probe-points can also be facilitated by "design for probe-ability" [13]. Access to signals can also be enhanced by, for example, creating areas larger than minimum metal width on critical signals such as buses, bringing critical signals to the upper metal levels with vias and leaving gaps in wide power supply buses.

7.5.1 FIB Deposition and Selective Material Removal

Historically, relatively crude laser ablation was used to locally remove dielectric to allow access for mechanical probes. FIB provides the improved spatial resolution and precision that are required for probe-point creation on submicron technology. Native FIB milling is not material selective and is capable of removing both metal and dielectric materials at a rate of ~0.1um^3/nC (cubic microns per nano-Coulomb) of ion dose. Gas injection enables a number of processes in addition to sputtering. A gas injection needle is positioned about ~0.5mm above the device surface to ensure maximum gas flux or partial pressure while minimizing the risk of device damage due to needle collisions. The gas is adsorbed onto the surface. The ion beam is used to drive a reaction in the adsorbed gas or a reaction between the adsorbed gas and the sample. The initial application of gas injection was metal deposition, which enables hole filling and thus the creation of probe-points and circuit interconnections. Dielectric deposition is also possible. Halogen-based gases provide material selectivity, increased material removal rates and dramatically improved overall FIB performance[14].

An organo-metallic precursor material that is volatile at or a little above room temperature is used for conductor deposition. The organic portion of the molecule is dissociated by the ion beam and exhausted through the vacuum system leaving the metal behind. Tungsten or platinum based precursor compounds are most commonly used and repeatably deposit amorphous material with a bulk resistivity of 50-200uΩcm. This is substantially higher than the bulk resistivity of an aluminum conductor in an IC but is adequate for probe-point creation and for re-routing CMOS logic signals.

A silicon oxide precursor similar to TEOS, (which is routinely used in bulk oxide deposition during IC manufacturing) is typically used for dielectric deposition. In some cases, it is combined with an oxidizing agent to oxidize the gallium ions that are inevitably trapped in the deposition. The deposited material is a very high resistance conductor with a bulk resistivity of ~10-100GΩcm and a breakdown field strength of ~10-100V/um. Dielectric deposition substantially increases the flexibility and the number of applications of FIB technology for probe-point creation and repair.

A halogen-based precursor containing iodine, bromine or chlorine (iodine is probably the most popular for the obvious safety and ease of handling reasons) is used to enhance metal milling rates resulting also in material selectivity, in this case, for aluminum. The halogen reacts with the aluminum to form a volatile compound, which is more readily removed by the gallium ion beam. Typically the aluminum removal rate is increased by a factor ~7-8X (versus ~2X for the surrounding dielectric) and redeposition of the milled material is virtually eliminated because the resulting molecules are volatile and thus tend not to stick to the device surface. They are also more readily removed by the vacuum system.

A fluorine-based precursor, Xenon Difluoride (XeF_2), is used for selective dielectric removal in much the same manner as for metal selective processes. Acceleration factors of 20-30X are commonly achieved with silicon oxide or nitride and up to 100-1000X with bulk silicon (a significant advantage for backside drilling - see the flip-chip section of this chapter).

7.5.2 End-point Detection

When milling a hole either to cut or contact a signal, accurate end-point detection is critical to assure clean cuts or good contact to a signal. Several end-point detection techniques are in use. FIB systems present the end-point detection signal as a waveform or image derived from one of several end-point signal sources available.

Transitions from one material to the next are shown as changes in the level of that signal. Gauging the exact point in time to stop milling is empirical.

Stage current is the simplest means of implementing FIB end-point detection beyond just imaging while milling. A pico-ammeter is connected between the sample stage and system ground. The stage current is the sum of the primary beam current, and secondary ion current, and secondary electron current into the sample. The primary beam current is constant. The secondary ion current is at least an order of magnitude smaller than the secondary electron current; consequently the stage current is the primary FIB beam current with the secondary electron current and variations superimposed.

The secondary electron signal can also be measured directly by integrating the video image signal over one or more frames. In principle, this direct measure of the secondary electron signal will be more sensitive than stage current since the beam current is not part of the signal. The disadvantage of this approach is that the gain of the secondary electron detector is superimposed on the end-point signal. Any changes in gain will result in substantial end-point signal changes. Using the secondary electron signal with windowing allows edge effects from the walls of the milled hole to be subtracted from the end-point signal, increasing its sensitivity. Edges result in preferential secondary electron generation and appear in the image (and end-point signal) as a bright constant signal that does not vary with changes in the material being milled at the bottom of the hole. Windowing software can be used to increase end-point detection sensitivity by only integrating the secondary electron signal from the center of the milled area. This approach is particularly useful when milling relatively high aspect ratio holes where end-point detection signal sensitivity becomes one of the primary limiting factors in determining the ultimate usable limit for precise milling.

Secondary ion current can also be used for end-point detection in much the same manner as the secondary electron signal as described above. For some material transitions, the secondary ion materials contrast change will be greater than the corresponding secondary electron change, and as a result secondary ion end-point detection would, in principle, be more sensitive. As the secondary ion signal is one to two orders of magnitude smaller than the secondary electron signal, this potential for increased sensitivity is usually more than offset by decreased end-point signal to noise ratio. Secondary ion end-point detection also has the advantage that it can be used in the presence of an electron flood beam as the electron flood beam swamps the secondary electron detector. SIMS (Secondary Ion Mass Spectrometry) can also be used to directly determine which types of ions are being milled. However, SIMS detectors are relatively uncommon on FIB systems used for probe-point creation and repair.

End-point detection techniques are substantially improved with halogen gas selective etching, which reduces material redeposition. Redeposition tends to mask material contrast changes at interfaces decreasing the overall sensitivity of end-point detection. Selective etching also results in satisfactory end-point signal from substantially deeper and higher aspect ratio holes for the same reason.

7.5.3 Probe-point Creation

There are several basic approaches to accessing conductors covered by a dielectric. Deprocessing or Global Depassivation can be used to expose conductors for probing. All of the layers of dielectric can be removed anisotropically with selective Reactive Ion Etching (RIE), leaving multiple layers of fully functional conductors standing on unetched dielectric bridges[15]. The advantage of this technique is that it is global and exposes all accessible signals in a single step. The primary disadvantage is the resulting changes in the effective dielectric constant in which the whole device operates. If the application is a subtle speed-sensitive problem, then global depassivation can easily mask or eliminate the symptoms completely. If the failure is defect related, global dielectric removal may also remove the defect itself (e.g. a particle embedded in the dielectric).

The FIB with dielectric preferential milling allows local depassivation by creating a "window" of local dielectric removal. The fact that only a window of dielectric is removed means that the device operation is less perturbed by dielectric constant changes. This approach also reduces the chance of inadvertently removing a defect. The top one or two metal layers can be depassivated in this manner. Milling selectivity is critical to the success of this operation. Selectivity is highest at relatively low beam current density, typically $<=\sim 10pA/\mu m^2$ depending on the gas flux. Even with a preferential milling capability, local depassivation is limited to two layers of metal by milling selectivity (~10:1 selectivity for dielectric:aluminum) since surface conductor damage becomes quite significant as the third conducting layer is uncovered. Adaptive masking of the beam can be used to mask out, or at least minimize, surface conductor milling and thus aid the preferential milling process[16].

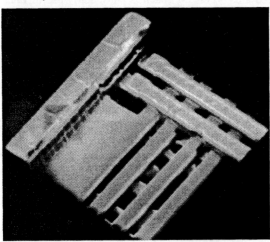

Figure 7.13. Advanced four-layer metal device with a local depassivation window is shown.

Since broad power buses frequently dominate top level metal, power-plane window creation followed by local depassivation of the underlying conductors is an important process. Without preferential milling capability, a power-plane window is an extremely challenging operation. FIB material removal rates are a strong function of landing angle. Increasing angles up to almost gracing incidence result in substantially increased material removal rates[17]. Any surface topography or material inhomogeniety will produce locally accelerated milling causing local pitting that usually results in non-uniform power-plane removal and damage to underlying conductors. If dielectric preferential milling is used to remove the overlying material and metal preferential milling is used on the power-plane itself, the material interfaces in effect act as barriers that slow the milling process and minimize or eliminate topography-induced pitting. Once the power-plane is removed, depassivating the underlying conductors is performed as described above. Figure 7.13 shows a power-plane window with underlying local depassivation.

In principle precision drilling is one of the simplest probe-point creation techniques. Since each node is exposed individually, it is more time consuming than the techniques identified above. In practice, therefore, it is most frequently used in situations where space and access are most limited. The most critical elements of precision drilling are hole aspect ratio and location accuracy. Figure 7.14 shows a cross-sectional diagram of a probe hole drilled with an unassisted or native FIB. The cross-section shows a characteristic ~6:1 depth to width aspect ratio caused by redeposition of milled material onto the sidewalls of the hole. Essentially, a dynamic equilibrium is achieved during milling between the rate of material removal and the redeposition rate. Halogen gas increases this aspect ratio to 15-20:1 as the milled material is a more volatile as halogen-based milled by products are more volatile and do not stick as readily to the side walls. For holes with aspect ratios greater than 1:1, the probe-point must be filled with conductor deposition even for e-beam probing. Without fill the secondary electron signal is distorted by the probe-point side walls and results in severe amplitude errors. The real challenge in the precise positioning of the hole is described in the discussion of overlay in the navigation tools section.

In addition to the four basic probe-point creation operations described, there is a whole range of combinations creatively devised by analysts to solve the specific problems.

7.5.4 Repair

FIB repair is now a standard tool in the development and debug of complex ICs. The ability to validate design changes is extremely valuable by facilitating multiple and proven design changes per mask design cycle. Once a particular set of modifications results in a fully functional prototype, the ability then to generate early engineering prototype samples significantly accelerates system development. A repair is essentially a series of cuts and conductive straps between probe-points. Dielectric deposition facilitates creation of the straps by allowing passivation of exposed metal, which is not to be contacted. The dielectric can also be used to cover deposited metal and reduce the length of subsequent metal depositions. The techniques described in the previous section apply equally to both probe-point creation and repair applications. Figure 7.15 shows a conductive strap. Note that the dielectric around the strap is dark. This indicates that deposition overspray, typical of all FIB conductor depositions, has been cleaned up usually with halogen assisted milling.

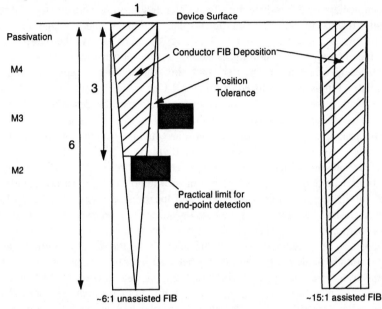

Figure 7.14. FIB hole aspect ratios for unassisted FIB v.s. halogen gas assisted are compared.

7.5.5 FIB Limitations

While FIB is an extremely powerful tool particularly for design debug applications, it has performance limitations. The native material removal rate is $\sim0.1\mu m^3/nC$ or $\sim0.1\mu m^3/nA$-s. With gas acceleration this can be increased by approximately an order of magnitude. Consequently FIB technology is very well suited to small operations but quickly becomes impractical if more than a few tens of cubic microns of material are to be removed. Laser enhanced etching and deposition address the reaction rate issue but lack the spatial resolution of FIB.

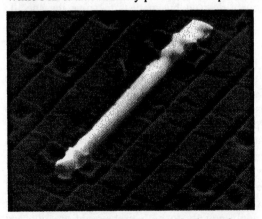

Figure 7.15. A typical FIB deposited metal strap is shown.

Conductor deposition

resistivity is ~200µΩcm which is two orders of magnitude higher than typical IC conductors. Long deposited lines can have very high resistance as a result. Deposition overspray results from interaction between the gas, the primary beam and the primary beam wings or tails, resulting in a substantial halo around deposited straps. The halo or overspray is not itself harmful for a single strap but can result in substantial leakage currents when two or more straps placed in close proximity. Overspray can be reduced by operating with low beam current or it can be removed by preferential metal milling.

High aspect ratio cuts and end-point detection limit the effective reliable working "depth" for a single FIB operation. Operating with low beam current provides cleaner cuts and easier end-point detection (especially through power planes). Cut isolation resistance is limited by conductive material redeposition and gallium implantation on the inner wall of the isolation cut. Isolation cuts should be wide and deep. High isolation resistance cuts, ~10-100MΩ, can be achieved with an unassisted FIB by making wider cuts and polishing sidewalls. Cuts with isolation resistance up to ~1TΩ can be achieved with halogen gases to address leakage from redeposition and gallium implantation[14]. Note that active region damage from gallium implantation can also influence leakage.

Electrical over-stress damage can also occur during the FIB process. Local charging of the device surface can result in damage due to sudden breakdown and discharge. Grounding all pins of the device & shielding insulating package surfaces with conductive tape reduces these charging effects. Floating conductors can occasionally, however, still charge, potentially damaging sensitive gate dielectrics. Gallium implantation, for the vast majority of devices, induces sufficient surface leakage current to eliminate this problem in most cases. A few devices, however, are still sensitive to damage from surface charging. This can be overcome either with a thin conductive carbon coating or by the use of a low energy electron flood gun. Avoiding the creation of floating gates through planning of the FIB operation sequence is also essential. FIB work should, in general follow the sequence of via contacts, deposition and isolation cuts in that order to avoid leaving floating conductors mid-sequence.

In addition to charging damage, ion-beam exposure can subtly change transistor parameters due to the gallium implantation. The use of halogen gases reduces damage by reducing the total ion dose required for a particular operation.

7.6 BACKSIDE PROBING FOR FLIP CHIP

Due to the increase in flip-chip utilization, productive flip-chip diagnostic tools and techniques are a critical industry need and active research area. Migration from traditional wire bonding with peripheral bond pads to high performance flip-chip packaging, with an array of solder bumps covering the front surface, is primarily driven by three factors: signal bandwidth limitations due to bond wire inductance, a few nH per mm, limits usable I/O bandwidth to approximately 300MHz. As pin counts continue to grow, peripheral pad spacing becomes a limiting factor. Actual die sizes have not increased significantly. Thus the peripheral area or space for bond pads does not increase. Bond pads sizes have not changed significantly. Flip-chip also helps to improve package form factors, heat dissipation and hence reliability.

Flip-chip is not the only driver of backside approaches for failure analysis. Even with wire bonding, conventional front-side signal access is increasingly limited by the growth in the number of layers of metal with the upper layers consisting almost exclusively of wide power busses. Global failure site isolation techniques from the backside have already been discussed in Chapters 5 and 6. The fact that silicon is transparent in the near Infra-Red, (IR) enables these global techniques as well as probing of individual circuit elements. As discussed previously, global techniques efficiently provide failure site isolation in many cases. However, many applications, such as design debug, require the ability to make electrical measurements. None of the global techniques provides the timing information that is essential for speed path measurements in design debug or for failure analysis of speed sensitive problems. Two backside approaches have been shown to be viable: backside drilling followed by mechanical or e-beam probing, and IR modulation-based optical beam probing.

7.6.1 Backside Drilling and Probing

FIB milling to probe signals from the backside is possible for nets, which can be exposed without affecting device functionality. This approach offers two advantages: real voltage measurements because of direct electrical signal access and the use of existing front-side probing tools and techniques. The disadvantages are that it is generally tedious, slow and provides access only to those signals, which exist in metallization accessible between active diffusion regions.

The device is first mechanically thinned to ~50-200μm using conventional mechanical grinding starting with a relatively rough grit and progressively reducing this to a fine diamond paste in order to generate a reasonably good, optically flat finish. Devices can in fact be successfully thinned to within 10μm of the active silicon surface. However, thicknesses less than 100μm substantially increase the chances of device damage due either to localized thermal stress or mechanical failure. The optically flat finish is required to capture a backside IR image for optical overlay-based navigation and also to help minimize topography induced pitting during FIB milling [18].

FIB milling is started in the approximate area for measurements using a maximum beam current with gas acceleration from XeF_2 (xenon difluoride). XeF_2 reacts with the bulk silicon to form volatile fluorides, accelerating the native FIB material removal rate by a factor of 100-1000X. The gas flux should be maximized. A precisely positioned hole, with an inverted pyramid profile and an effective aspect ratio (total depth to width at top) of ~1:1 is drilled between the active regions to expose metal-1 or polysilicon conductors for probing. The low aspect ratio facilitates probing without the need for extraordinary probe-point creation techniques. The same techniques can also be used for flip-chip prototype repair when front-side signal access is impractical.

Precise positioning is achieved by using a backside IR optical image superimposed or overlaid on the FIB image. Final alignment of the overlay with the FIB image is achieved by using the FIB milled hole itself (which is visible in both the IR optical and FIB images) for precise registration of the overlay. Measurements can then be made in the usual manner with either an e-beam or mechanical probe.

While this process is clearly very limited, slow and tedious, it is the only way to get real DC or quantitative AC voltage amplitude measurements from flip-chip packaged ICs (as IR modulation-based probing provides only a relative signal).

7.6.2 IR Modulation-Based Probing

IR modulation-based probing uses an IR optical system to deliver a pulsed IR laser beam probe onto the active region of transistors. The signal is derived by detecting subtle modulation of the reflected IR beam. Figure 7.16 shows a typical setup for IR modulation-based probing.

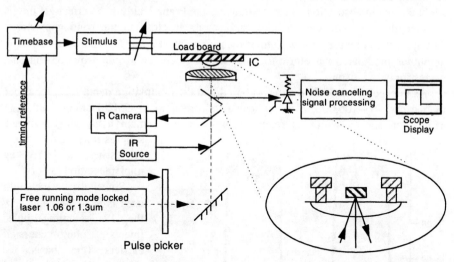

Figure 7.16 Schematic of the optical setup for IR modulation based probing is shown.

As the IR beam passes through the diffusion region of the transistor, three types of modulation can occur depending on the geometry and material involved, and the current and voltage state of the device being probed. These are specifically: amplitude modulation[19], phase shift or propagation delay[20] and polarization change[21].

The bandwidth for measurement of the modulations is limited by the laser pulse width and stimulus source jitter stability. Tester or stimulus source jitter is increasingly the limiting factor in sampling-based risetime measurements of all kinds. Typical laser pulse widths and jitter yields ~50ps measurement bandwidth. Thus, this technique is capable of making the precise timing and propagation delay measurements required for design debug.

Polarization based modulation only works in optically active materials like GaAs. It has relatively limited applicability for super highspeed GaAs. Phase shift based modulation detects changes in active region refractive index resulting from charge carrier density changes, which are most prevalent in bipolar or current mode logic. The refractive index change results in subtle beam propagation delays, perhaps better described as optical phase shifts. This delay can be measured interferometrically by passing a reference beam through a neighboring non-active area of the silicon. This also cancels errors from thermal modulation effects. The primary limitation of this technique is that the modulation is very weak in conventional CMOS devices since the carrier density changes by a relatively small amount as CMOS circuits switch.

Amplitude-based modulation uses changes in the electric field across a PN junction to make measurements. The electric field across a reverse biased junction results in a subtle shift in the effective band gap energy, which in turn produces a slight change in absorption. This is known as the Franz-Keldysh[22,23] effect. If the IR wavelength or photon energy used for probing is close to the band gap energy ~1.06μm, it can be used for waveform measurement. Figure 7.17 shows an example amplitude modulated waveform together with a reference acquired by conventional means for comparison.

As this technique is field strength based and the modulation depth is on the order of one part in ~100,000, it is applicable to conventional CMOS circuits although acquisition speeds with this relatively weak signal remain a challenge. Noise suppression and signal processing are clearly key implementation factors.

Since the wavelength used is near the band gap, the IR sampling pulse injects photo-generated carriers (the basis for OBIC and LIVA based fault isolation) at the instant of sampling. During the steady state between transitions, this is of little consequence; however, during switching the potential for the measurement to be somewhat invasive exists. Acceptable results can be repeatably achieved with

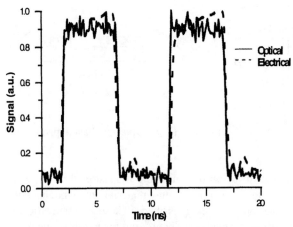

Figure 7.17. Waveform acquired using IR amplitude-based modulation probing is compared to waveform acquired using e-beam probing. The IR waveform was acquired in ~15 minutes with a 10us test pattern loop length.

moderate laser power balanced with good signal processing to achieve reasonable waveform acquisition speeds.

The use of a customized IR confocal optical system, which only collects light from a single in-focus optical plane, has two advantages. The first is higher image contrast, which partially offsets the wavelength limited IR resolution. The second is a reduction in noise from unmodulated out-of-focus planes (Note that usable IR modulation only occurs in the high field strength junction region).

Improvements in optical resolution may be possible using immersion objective lenses. In addition, photons falling outside the immediate modulation volume do not result in errors, just a weaker signal. In comparison, electrons falling on the edge of conductors during e-beam probing can result in severe amplitude errors due to preferential secondary electron generation from the edges.

Precise time resolution is achieved by using a mode locked laser as a source of short pulses for equivalent time sampling in much the same way as an e-beam prober does. Pulse widths shorter than 50ps are readily achieved with these lasers. With advanced laser pulse compression techniques, substantially shorter pulse widths can be achieved. Most commercial testers have jitter specifications ~10ps RMS, the stimulus source, rather than the sampling pulse, becomes the primary measurement bandwidth limiting factor in real IC's.

Mode locked lasers are fixed frequency, free running (~50-200MHz) oscillators, and consequently synchronization of the sampling pulses to the test pattern cannot be achieved with a simple trigger signal as is the case with e-beam probing. The simplest solution is to use the timebase, effectively in reverse, to delay the start of the test pattern as is shown in Figure 7.7, rather than to delay the beam pulse itself. An additional delayed timing signal is required to drive a highspeed laser pulse picker (or fast electro-optic switch) which selects a single laser pulse from the continuous stream of pulses.

While in principle this approach to synchronization seems straightforward, some high performance testers rely on the stability of internal oscillators for overall tester timing accuracy. Frequent phase changes impact oscillator stability reducing overall timing accuracy and increasing jitter. A more robust solution is of course desirable.

It should also be noted that unlike e-beam probing, the mode locked laser pulse width is constant and thus acquisition speed is constant for different time per division settings. For e-beam probing the pulse width is scaled to match the display time per division and thus acquisition speed increases as the time per division (or pulse width) is increased.

One other potential IR flip-chip time resolved probing technology deserves a mention here. Emission microscopy, both front-side and back-side is covered in chapter 6. A potential extension to this approach called PICA (Picosecond Integrated Circuit Analysis) was recently published[24]. MOS transistors when switching generate weak, broad spectrum light pulses due to hot carrier injection. The IR component of these pulses passes through a thinned silicon substrate and can be integrated by a very sensitive, high resolution, time-resolved camera. These pulses, that essentially carry current information, can then be interpreted and used to measure rise-time and propagation delay information.

7.6.3 Application Tips for IR Modulation-Based Probing

Sample preparation is similar to other IR-based techniques. Since IR transmission through the silicon substrate decreases exponentially with thickness, thinning is very beneficial provided mechanical integrity and thermal dissipation are not sacrificed. Anti-reflection coatings are also beneficial for all IR-based diagnostic techniques. The use of an anti-reflection coating on the polished substrate can increase the reflected IR power by up to 50% due to the optical impedance miss-match between the refractive index of air, n=1.0, and the relatively high bulk silicon IR refractive index of n=~3.5.

The majority of heat generated by ICs escapes through the substrate due to the relatively good thermal transmission characteristics of bulk silicon. The thermal mass for dissipating power is significantly reduced by thinning the silicon and removing the heat sink if one is attached to the back of the device. Various cooling techniques may have to be employed during probing.

Probe-points: As CD's (Critical Dimensions) continue to shrink, it may be appropriate to consider designing in reverse biased PN junctions as probe-points on key signals. Today many manufacturers place antenna protection diodes on long bus lines to protect against electrostatic charge build up during the manufacturing process. If available these make good test points.

ACKNOWLEDGEMENT

Photographs courtesy of Schlumberger ATE.

REFERENCES

1 Brown. Fast, Reliable Mechanical Probing of Submicron Features - A New Tool That Combines Gas-Assisted FIB and Mechanical Probing. European Optical Beam Testing Conference, 1995.
2 Smith K. PhD Dissertation, University of Cambridge, England, 1956.
3 Menzel E, Kubalek E. Fundamentals of Electron Beam Testing of Integrated Circuits. SCANNING, 1983, 5, 103.
4 Gopinath A, Sanger C. A Technique for Linearization of Voltage Contrast in the SEM. J. Phys. E: Scientific Instruments, 1971, 4, 334.
5 Kruit P, Read EH. J. Phys. E: Scientific Instruments, 1983, 16, 313.
6 Takahashi H, Masnaghetti D, Richardson N. "Double Gated Integrator Scheme for Electron Beam Tester" US Patent #5,210,487, May 11, 1993.
7 May P, Halbout JM, Chiu G. Non-contact Highspeed Waveform Measurements with Picosecond Photoelectron Scanning Electron Microscope. IEEE Journal of Quantum Electronics, 1988, 24 (2), 234.
8 May T, Scott G, Meieran E, Winer P, Rao V. Dynamic Fault Imaging of VLSI Random Logic Devices. Proceedings International Reliability Physics Symposium, 1984, 95.
9 Wada SI, Nakamura T. Automatic Fault Tracing Using an E-beam Tester with Reference to a Good Sample. Proceedings International Symposium for Testing and Failure Analysis, 1997, 243.
10 Concina S, Liu G, Lattanzi L, Reyfman S, Richardson N. Software Integration in a Workstation-based E-Beam Tester. Proceedings International Test Conference, 1986, 644.

11 Talbot C, Masnaghetti D. Large Area Overlay for Accurate Location of Invisible Buried Conductors. Ion Microbeams - Generation & Applications, Germany, May 1994.

12 McCluskey EJ. Built-in Self-Test Techniques. IEEE Design and Test of Computers, 1985, 2 (2), 21.

13 Lee W. Engineering A Device For Electron-Beam Probing. IEEE Design and Test of Computers, June 1989.

14 Ximen H, Talbot CG. Halogen-Based Selective FIB Milling for IC Probe-Point Creation and Repair. Proceedings International Symposium for Testing and Failure Analysis, 1994.

15 Baerg W, Rao VRM, Livengood R. Selective Removal of Dielectrics from ICs for Electron-Beam Probing, IEEE/ International Reliability Physics Symposium, 1992.

16 Talbot CG, Masnaghetti D, Ximen H. "Self-Masked FIB Milling", US Patent #5,616,921 April 1, 1997.

17 Wilson IH, Chereckdjian S, Webb RP. On the Variation of Sputtering Yield with Angle of Ion Incidence. Nuclear Instruments and Methods in Physics Research, 1985, B7/8, 735.

18 Livengood RH, Rao VR. FIB techniques to Debug Flip-Chip IC's. Semiconductor International, March 1998, 111.

19 Frova A, Handler P, Germano FA, Aspnes DE. Electro-Absorption Effects at the Band Edges of Silicon & Germanium. Physical Review, May 1966, 145 (2).

20 Keller U, Diamond SK, Auld BA, Bloom DM. Non-Invasive Optical Probe Of Free Charge And Applied Voltage In GaAs Devices. Applied Physics Letters, 1987, 53, 388.

21 Heinrich HK. Picosecond Noninvasive Optical Detection of Internal Electrical Signals in Flip Chip Mounted Silicon IC's. IBM Journal of Research and Development, 1990, 34, 162.

22 Franz W. Naturforsch., 1958, 13a, 484. &

23 Keldysh LV. Soviet Physics JETP, 1958, 7, 788.

24 Kash JA, Tsang JC. Full Chip Optical Imaging of Logic State Evolution in CMOS Circuits. IEDM 1996, Late News Paper.

8

IC DEPROCESSING

Daniel Yim
Advanced Micro Devices

Deprocessing (also called delayering), as the name implies, is the systematic process of removing the thin film layers of the die after it has been exposed or removed from the package. The purpose is to provide the analyst more visibility and accessibility to areas below the surface of the die where much of the electrical activity takes place. Accessibility is typically defined in terms of electrical access to signals for failure site isolation. On the other hand, visibility provides access to defects for physical and chemical characterization. Therefore, deprocessing is a critical step in the failure analysis process since improperly or hastily implemented deprocessing procedures can result in the loss of key pieces of information needed to understand the physical cause of failure. Errors as a result of poor deprocessing are among the most difficult in failure analysis to overcome. Deprocessing can be performed on a die while still in a package or on a die removed from the package with fundamentally the same methods. When deprocessing in the package, the package serves as a convenient holder for the die and the extra step of removing the die from the package is avoided. Dice removed from the package are typically attached to a glass slide with wax or epoxy. In this form, influences from the packaging materials during the delayering process are reduced. The same methods also apply to the deprocessing of wafers or wafer fragments. Deprocessing can be performed selectively as well as globally depending on the nature of the analysis required. Specialized diagnostic tools discussed in Chapter 7, such as the focus ion beam (FIB) and laser etcher are available to provide selective localized deprocessing capability. Due to the increasingly high number of interconnect layers and increasingly small feature sizes, deprocessing has become quite a complex process requiring high degrees of skill. In practice, typical deprocessing of the die is accomplished in discrete steps with requirements for inspection (see chapter 10) between each step of the area to which a defect has been isolated.

8.1 IC WAFER FABRICATION

Deprocessing can be viewed as a reversal of the wafer fab process. Thin film layers are removed sequentially in reverse order of application in the wafer fab. As such, an understanding of the wafer fab process is a critical element of deprocessing. The following discussion on silicon IC wafer fabrication is generic and limited to the basic concepts associated with CMOS (Complimentary Metal Oxide Semiconductor Field Effect Transistor) technology, the most commonly used transistor technology.

Basic silicon IC wafer fabrication begins with the selection of starting material. Blank, single crystal silicon wafers or substrates have a variety of properties, which can significantly influence the etching of the silicon deprocessing and delineation. These include crystal orientation, doping type (p or n), concentration of doping, and presence of an epitaxial layer. For example, heavily doped material tends to etch more quickly than one that is less doped, and n-type more so than p-type. Etch rates and patterns are also affected by the crystal orientation.

The formation of active devices such as transistors and diodes, and passive devices such as resistors and capacitors, is commonly referred to as the "front end" of the wafer fab process. The required patterned structures are accomplished through a photosensitive process called photolithography. It is the selective exposure of photoresist, a photosensitive film, which defines the areas to be processed or protected from processing, such as an etching or an implantation step. Selective exposure is achieved by exposing the photoresist through a photomask. The photoresist is not retained as part of the finished IC, providing only the pattern for a particular process step.

CMOS technology requires both n-channel and p-channel transistors. Since only one type of transistor can be formed directly in the substrate, a "well" of opposite doping must be created, such as an n-well in p-type starting substrate for the p-channel transistors. The p-type substrate is typically lightly doped and is based on providing a background concentration that will ultimately provide good n-channel transistor characteristics, but is not so high as to require an overly high concentration of n-dopant to overcome the p-dopant substrate concentration when creating the n-well. However, the low voltage threshold and high depletion width spreading created by the use of a lightly doped p-type substrate allows for parasitic NMOS channel formation as well potential "punch through" problems. To mitigate possible channel formation in the substrate, additional boron implants or "twin-well" processes, which support formation of both types of transistors, are frequently employed. Typically, the pattern for well formation is defined in a thick thermally grown oxide of about 1000Å. This oxide may eventually contribute to the isolation between devices to prevent unwanted or parasitic electrical paths from forming between devices, but often is removed prior to field implant followed by actual field oxide isolation formation. Implant of n-dopant is used to form the n-well, and in a "twin well" process, a p-dopant implant is performed similarly. Transistor formation is most commonly performed using a self-aligned gate process. A thermal oxide is grown providing the dielectric that creates the transistor gate. Polycrystalline silicon (polysilicon or poly) is then deposited on the gate dielectric by chemical vapor deposition, possibly doped by implantation, and patterned[i]. Polysilicon offers good electrical and physical characteristics such as work function and thermal matching to the silicon substrate. The polysilicon is used as the transistor gate electrode and helps define the formation of the source and drain junctions surrounding the gate. In order to create the exact source to drain spacing and structure required, sidewall oxides are commonly added prior to all or part of the junction implantation. Separate dopant implants are performed to create the source and drain junctions of the n-channel and p-channel transistors. Once the transistors are formed, the entire structure is covered with an oxide.

Upon completion of the CMOS transistors, interconnect layers are required to connect the transistors together. The gate polysilicon commonly provides a first level of interconnection. Although intrinsic polysilicon is hundreds of thousands or even megohms per square, by using dopants such as those to form junctions in the silicon, useful resistivity can be achieved on the order of tens of ohms per square. The interconnection processes beyond polysilicon are commonly referred to as the "backend" of the wafer fab process. Since the inception of IC manufacturing, the most widely used material for IC interconnections is aluminum metal, sometimes in combination with barrier/adhesion layers. Aluminum metallization provides low ohmic contact to highly doped n+ and p+ silicon and meets the interconnect requirements for a very wide range of applications for both MOS and bipolar designs[1]. Before metal is deposited, contact holes are patterned in the oxide protecting polysilicon, thus selectively controlling where the metal is to make contact with the layer below. Metallization is then deposited and subsequently patterned using another photolithography step. An oxide, commonly referred to as "interlevel dielectric", is deposited to electrically isolate the underlying metallization structure, preventing the metal lines from inadvertently shorting together due to the presence of defects and other unwanted contamination. While the most commonly used interlevel dielectrics are intrinsically oxide in nature, the properties of the interlevel dielectric vary drastically with the process by which it is applied, the addition of dopants, and subsequent heat treatments. For multilevel metallization processes, this four-step process, creating contacts, metal deposition, metal patterning, and dielectric deposition, is then repeated again. The contact holes connecting metal to metal are referred to as vias.

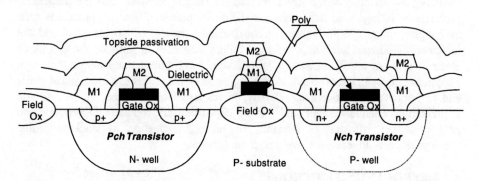

Figure 8.1. Basic CMOS Process is shown in cross section.

Finally, the entire structure is covered with a non-conducting layer, called passivation, to prevent physical damage as well as the penetration of moisture and other contaminants. The final masking process creates relatively large openings in the passivation to allow probing and bonding, completing the IC fabrication process. Passivation is usually composed of layers of oxide, nitride, polyimide, or some combination thereof.

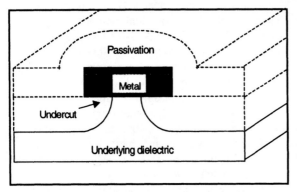

Figure 8.2. Undercutting of metal during passivation removal is illustrated. (Figure from W. Baerg, V.R.M. Rao and R. Livengood, "Selective Removal of Dielectrics from Integrated Circuits for Electron Beam Probing. Proceedings International Reliability Physics Symposium, 1992, 320 ©1992 IEEE.)

The processes described above and the resulting structure illustrated in figure 8.1 is a very simplified, but typical example of CMOS processing. State-of-the-art processes in the sub micron range incorporating gate oxides under 200Å include many enhancements and refinements to improve device performance and reliability. Because today's circuits places great emphasis on transistor speed and speed between logic blocks, device junctions are fabricated with sophisticated, multi-implant schemes to decrease the effects of junction capacitance, the effects of depletion layer spreading, and reduce hot carrier effects. The development of refractory silicides improves upon the best polysilicon line resistance by another order of magnitude. With decreasing feature sizes, the shrinking aspect ratio of the holes formed for making contacts and the associated problems in filling those holes required the development of separate processes such as Tungsten plugs. With higher current demands, the use of barrier metal and the inclusion of dopants such as copper into aluminum help to mitigate the effects of electromigration. Planarization processes have been added to improve photolithography in the deep submicron region as well as to relieve mechanical stress and to improve step coverage. Finally, lower temperature processes such as ion implantation, chemical vapor and plasma deposition are routinely incorporated to help minimize diffusion of dopants, impurities, nucleation of stacking faults, precipitates, and dimensional changes in the substrate[2].

8.2 DEPROCESSING METHODS

Deprocessing or delayering usually implies the removal of an entire layer across an IC. However, with more sophisticated analysis, it can be useful to perform selective deprocessing to supplement global deprocessing procedures. There are three basic methods for global removal of layers; 1) wet etch, 2) plasma etch, and 3) mechanical polish. These techniques are not mutually exclusive. For example, many deprocessing schemes include plasma etch of dielectrics and wet etch of metals. Similarly, mechanical polishing is commonly combined with chemical deprocessing techniques. In many cases, the same techniques and tools developed for etching materials during wafer fabrication can be adapted for failure analysis deprocessing. The etchants used in the wafer fab provide, at a minimum, an excellent starting place for development of deprocessing etchants.

8.2.1 Wet Chemical Deprocessing

Wet chemical etching is the oldest method for IC deprocessing and is still frequently used. Control of etch rates is primarily accomplished by temperature, time, and concentration of the etchant solution. In many cases, good selectivity is desired so the etchant is selected based on its ability to etch the desired material while minimizing damage to the underlying layer and other critical features. Wet chemical etching, when used for failure analysis, is normally a global etch technique although photoresist masking has been employed to etch out only selective areas. Without selective masking, wet chemical removal of material occurs across the entire IC. The primary issues with chemical wet etching are selectivity and its isotropic nature. Isotropic etching produces unwanted undercutting such as those shown in figure 8.2. As a result of undercutting, narrow metal lines will have a tendency to lift off the surface[3]. Isotropic etching also leads to etching of materials through holes such as vias, known as "etch through" shown in figure 8.3. Thus removal of upper metallization layers can result in unwanted etching of underlying metallization through the vias. However, with the advent of via plug technology, this effect has been greatly reduced.

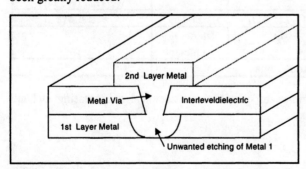

Figure 8.3. Etch through of metallization is illustrated.

The problems associated with selectivity to metals and isotropic etching have generally limited the use of wet etching of silicon oxides to devices with large feature sizes. Silicon nitride etching is similarly limited, but in addition, nitride etches have been relatively limited in effectiveness. Metallization wet etching is still commonly performed on a broad range of devices. Selectivity of metal etches to most dielectrics is very good and etch-through is frequently halted by barrier/adhesion layers and tungsten plugs in vias, making it less of an issue during deprocessing. Silicon etching and crystal defect decoration are still commonly performed using wet etches. The applications of wet chemical etching are summarized in table 1.

Common Materials	Common Etchants	Selectivity	Comments
Silicon	HF with oxidizers such as HNO_3 or Chromic Acid		Most Commonly Used Very Caustic Chrome is a Heavy Metal
Silicon oxides	HF buffered with NH_4F	Moderate to Metals	
	10:1 HF	Moderate	Attacks Silicon if left too long
Silicon Nitride	Nitrietch[5]	High to metals, low for CVD oxides	45-50C sensitive to temp
	HPO_3	Poor	Caustic, etches Al
	Buffered HF	Poor	Etches oxides
Polysilicon	HF / acetic acid / HNO_3 (Dash)	Moderate	Attacks oxides and silicon if left too long
Aluminum	HPO_3 / HNO_3 / acetic acid	Good	Must be heated ~70C
	H_2O_2 / H_2SO_4 (Piranha)	Poor	Used for stripping, Very reactive, attacks W, Ti, TiW, TiN
W, Ti, TiW, TiN	H_2O_2 / H_2SO_4	Poor	See above
	H_2O_2	Moderate to metals	
Gold wire bonds	Aqua regia		

Table 8.1. This table summarizes common wet chemical deprocessing etchants. Many variations on basics exist.

8.2.2 Plasma Deprocessing

Plasma etching, also known as dry etching, is accomplished in a plasma reactor. An evacuated and pumped chamber is continuously backfilled with a pure gas or mixture of gases. The plasma is generated when sufficient energy is applied across two electrodes to form neutral free radicals and charged ions from the gases in the chamber. The free radicals are not affected by the electric field set up within the chamber since they are neutral, but they do diffuse and move about randomly.

Etching occurs when the radicals strike the sample, reacting with the material on the surface to form volatile by-products that are then pumped out of the chamber[4]. There are three main types of plasma etchers: the barrel etcher, the parallel plate etcher, and the reactive ion etcher.

The barrel etcher (see figure 8.4), the first available dry etcher, is the simplest of the plasma etching systems consisting of a quartz chamber and two external RF electrodes. Early barrel etchers provided an effective mechanism for removing silicon nitride passivation films since compressive forms of nitride were particularly difficult to wet etch. Albeit slower than wet etch, with the proper gas mixture of CF_4 and H_2[5], these systems provided acceptable removal of oxides with good selectivity to metal and silicon. Oxides are etched by the CF_3 radical, which is very reactive to oxides, and inclusion of H_2 increases formation of CF_3 from the CF_4 mixture in the chamber. In reality however, the barrel etcher etches oxide very slowly because the short-lived CF_3[3] is produced near the electrodes where the electric field is highest, requiring the CF_3 to diffuse a great distance to the sample. For etching nitrides, O_2 is added to the CF_4 to reduce unwanted etching of oxides and silicon. The atomic fluorine free radicals from CF_4 react with silicon nitride to create volatile fluorides. SiF_6 is another source of fluorine radicals, but is very corrosive placing harsh demands on the system[6]. The barrel etcher removes nitride films quickly with good selectivity since there is minimal etching of the underlying oxide as well as to metal. A major drawback of this system is a lack of temperature control causing both etch rate and fluorocarbon polymer formation to increase as the sample gets hotter over time. Additionally, barrel etching is primarily isotropic meaning the effects of undercutting have not been alleviated.

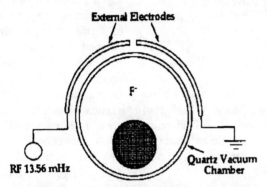

Figure 8.4. Basic elements of the barrel etcher are shown. (Figure from W. Baerg, V.R.M. Rao and R. Livengood, "Selective Removal of Dielectrics from Integrated Circuits for Electron Beam Probing. Proceedings International Reliability Physics Symposium, 1992, 320 ©1992 IEEE.)

The parallel plate etcher (see figure 8.5) improves upon the barrel etcher by providing temperature control for added stability. In addition, the sample is placed on the grounded electrode since parallel plate electrodes are placed at the top and bottom, inside the vacuum chamber. With this configuration, the system can take advantage of the ions created in the plasma by creating a steady state electric field supplied by a low frequency (450KHz) RF to the top electrode through a capacitor to

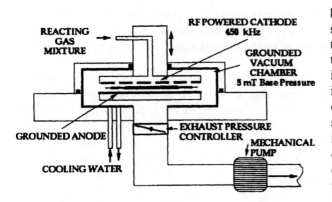

Figure 8.5. The basic elements of the parallel plate etcher are shown. (Figure from W. Baerg, V.R.M. Rao and R. Livengood, "Selective Removal of Dielectrics from Integrated Circuits for Electron Beam Probing. Proceedings International Reliability Physics Symposium, 1992, 320 ©1992 IEEE.)

block DC current[3]. The small DC bias between the plates created by the RF power source is insufficient to initiate ion movement on its own. Thus, the RF signal is also important in initiating movement of the negatively charged ions towards the positively charged anode at the chamber bottom due to the high plasma pressure (i.e. short free mean path). With directional ion bombardment realized (see figure 8.6), polymer formation is now restricted to features parallel only to the direction of ion movement, such as the sidewalls, providing the anisotropic etching characteristics. With the parallel plate configuration, CF_3 radicals produced near the electrodes, which were too far away from the sample in the barrel etcher, are now close enough to contribute to etching oxide at a fairly high rate. However, with bombardment occurring on the top electrode as well, sputtered metal from the top electrode can now deposit on the samples under etch, inhibiting etching as it accumulates [3,4].

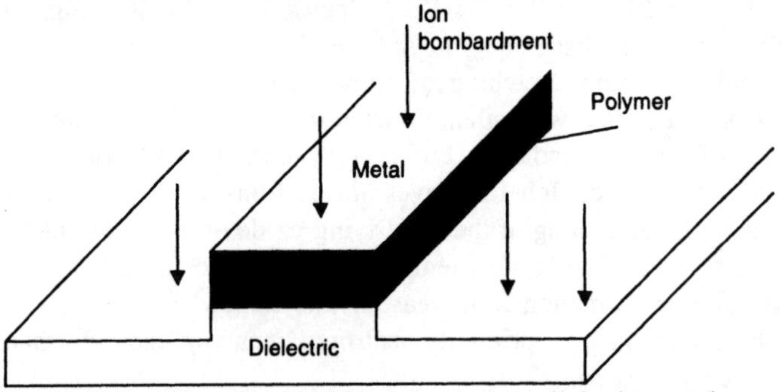

The last etcher type is known as a reactive ion etcher (RIE). There are three subtle, but basic differences from the parallel plate etcher, which improves upon performance. First, the chamber is under much lower operating pressures

Figure 8.6. Directional etching in the parallel plate etcher results in the desired anisotropy. (Figure from W. Baerg, V.R.M. Rao and R. Livengood, "Selective Removal of Dielectrics from Integrated Circuits for Electron Beam Probing. Proceedings International Reliability Physics Symposium, 1992, 320 ©1992 IEEE.)

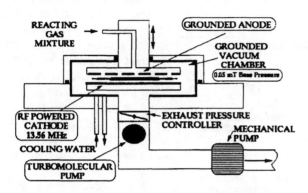

Figure 8.7. The basic elements of the RIE are shown. (Figure from W. Baerg, V.R.M. Rao and R. Livengood, "Selective Removal of Dielectrics from Integrated Circuits for Electron Beam Probing. Proceedings International Reliability Physics Symposium, 1992, 320 ©1992 IEEE.)

achieved by using a turbomolecular pump. Operating pressure is about 10 millipascal compared to 100-500 millipascal for parallel plate etching allowing the reaction by-products to be pumped away much more quickly and, therefore, less likely to redeposit upon the sample. Second, a much higher frequency (13.56MHz) RF source is used to supply the energy to create the plasma. The higher frequency provides a larger electric field between the two plates. Third, the grounded anode is designed with a surface area approximately 2x larger than the cathode and their positions are swapped so that the sample now sits upon the cathode at the bottom of the chamber. This asymmetry in electrode area creates a larger potential drop across the capacitance between the lower cathode and the plasma than the upper anode to the plasma resulting in more ion bombardment to the sample on the lower plate and less sputtering of the metal anode at the top[3]. These system improvements provide the user enhanced control over selectivity and anisotropy, resulting in greater reproducibility. When metal etching, compounds of chlorine instead of fluorine are used in the RIE. The gases are not interchangeable. As with dielectrics, chlorine based RIE's provide greater control for removing aluminum. Upper layer aluminum metal is removed uniformly with no impact to the underlying metal layer as

Figure 8.8. Sample with three levels of metal exposed by RIE is shown.

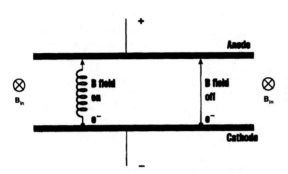

Figure 8.9. Schematic of MERIE is shown. From THE SCIENCE AND ENGINEERING OF MICROELECTRONICS FABRICATION by Stephen A. Campbell. Copyright © 1996 by Stephen A. Campbell. Used by permission of Oxford University Press, Inc.

typically observed during wet etching. However, with the incorporation of via tungsten plug technology to connect the different levels of metal, metal RIE using highly toxic chlorine is no longer a necessity. Wet metal etching is much more convenient and accessible to use, and as previously mentioned, does not effect the tungsten plug.

When etching using RIE, the unwanted formation of "RIE grass" can be observed. It is an etching artifact caused by the redeposition in small patches of sputtered materials onto the sample. These patches are inert to the chemically active species in the plasma[4,6]. Since the material is nonreactive, it effectively acts as a micromask preventing the underlying film from etching evenly. RIE micromasking can occur in several ways. Aluminum grass is caused by the redeposition of sputtered IC metallization exposed during ion bombardment in the presence of oxygen and fluorine. Polymer grass is a normal by-product of the standard plasma etch chemistry mentioned earlier. Adding oxygen into the system or reducing the gas pressure minimizes its formation. Other types of redeposition occur when IC package material, such as gold, is not well protected during RIE and redeposits on the sample surface[7].

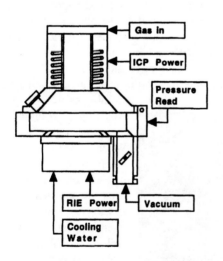

Figure 8.10. Basic elements of the ICP are shown[6]. From 22nd International Symposium for Testing and Failure Analysis, (1996), ASM International, Metals Park, OH 44073-0002, p 74 (Fig. 1).

More recent advances in RIE technology have included Magnetically Enhanced RIE (MERIE) and Inductively Coupled Plasma (ICP). MERIE uses magnetic fields perpendicular to the direction of ion travel[8]. The magnetic field induces the electrons to travel in helical paths, greatly increasing the ionization probability. A factor of 50 increase in ion generation in the plasma is quite achievable (see figure 9). As a result, MERIE produces a much higher ion flux at much lower energies than traditional RIE and ion bombardment damage of the sample is significantly reduced[9]. ICP utilizes two RF power sources: one to generate the plasma and one to generate the electric field (see figure 10). The ICP source is the higher RF energy source used for plasma generation in the upper half of the chamber away from the sample. Since it does not electrically affect the sample, control can be more precise providing large amounts of power. The second power source is a typical RIE source using less power, creating the electric field at the sample for control of the anisotropic etch[7].

8.2.3 Mechanical Polishing

The third method for deprocessing is mechanical polishing, also commonly referred to as parallel polishing or lapping. It is the physical deprocessing of an IC through the use of abrasive materials. The process is time consuming and meticulous by nature. However, with increasing layers of thin films, it has become more competitive with chemical deprocessing in terms of overall deprocessing time. As the number of required deprocessing steps increase, the number of deprocessing induced artifacts increases, making mechanical polishing more attractive. In addition, increasingly smaller feature sizes amplify the probability of unwanted metal lifting during dry and wet chemical deprocessing. A major drawback of lapping is the difficulty in maintaining planarity of the surface, especially over a large region. However, if the defective area of the die has been limited by failure site isolation techniques, mechanical polishing is an effective alternative to chemical deprocessing. It is performed using equipment and abrasives readily available in most failure analysis laboratories for cross section preparation (see Chapter 9).

From an inspection perspective, the advantages of using polishing over wet and dry etching techniques is the ability to stop mid layer and view features in the area of interest within the same plane. With wet and dry etching techniques, it is not possible to view, for example, metal and oxide at the same plane because removal of oxide would leave only the topography of the metal. Mechanical polishing is especially useful on multilayer interconnect processes with advanced planarization techniques. With a very

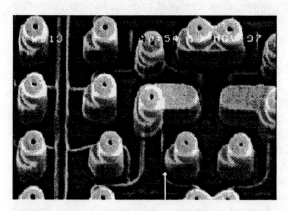

Figure 8.11. Tungsten plugs exposed by mechanical lapping are shown in SEM image.

planarized process, it is often difficult to discern critical features when viewed from the top because of the lack of depth perception. Polishing allows for visual inspection of protruding processing defects such as short causing stringers, filaments, and particles that might otherwise be loss during a wet or dry etch procedure.

Polishing for deprocessing is somewhat of an art form with a variety of custom equipment setups and procedures developed through trial and error. In general, the die is always removed from the package. Procedures have been developed for polishing the die alone or mounted in molding resin. In either case, careful alignment of the sample is essential to optimize planarity, uniformity, and angle of attack. Pre-characterization work to determine polishing rates as a function of the thin film materials and abrasives before deprocessing an actual sample is essential.

More recently, mechanical polishing for die access has garnered interest because of the increasing use of Flip Chip and some Chip Scale Packages (CSP). These packaging techniques do not allow easy access to the front surface of the IC. As a result, new failure analysis approaches have been developed from the backside (See Chapter 4).

8.3 NEW CHALLENGES

Until recently, the materials discussed in this chapter have proven electrically sufficient to sustain the performance and functionality improvements, essential to the IC business. In order to sustain those improvements, especially in the face of ever shrinking feature sizes, a new generation of materials are being developed and characterized. These materials exhibit improved electrical performance to withstand the higher speeds generated by higher current densities, lower capacitances and resistances, and higher electric fields. In addition to having the right electrical properties, the material must be sufficient to meet the mechanical demands of adhesion and stress as well, all while retaining the reliability robustness that's come to be expected by customers. However, their chemical and mechanical properties are, in general, not well understood at the feature sizes at which they will be employed. New deprocessing techniques will have to be developed to address the new materials. In addition to issues associated with removal of these materials, deprocessing approaches must facilitate observation or detection of any new failure mechanisms that may occur with these materials.

High K dielectric materials are being introduced to alleviate the problems associated with the thinning of gate oxide dielectrics which are now measured in monolayers. Similarly, the sheet resistivity of aluminum is is proving inadequate for high-speed, deep submicron devices. Copper is already replacing aluminum. Experts predict, however, that copper will probably be sufficient only for the next several generations of process technologies and must be eventually replaced. New deprocessing procedures will have to be developed for removal of copper, and new safety issues considered with the disposal of heavy metal copper in the lab since traditional etchant disposal methods will not suffice. These very same issues exist for low K dielectrics, which will be employed between metal layers to reduce interlevel metal capacitance. From an FA perspective, these new materials pose new challenges since little is known about any new failure mechanisms that could be created and only little literature exists on the removal and etching of these layers at this time. Lastly, as mentioned previously, new package technologies could profoundly affect the ways the analyst performs deprocessing. Already, the deployment of Flip Chip and Chip Scale Packages has dictated a greater reliance on polishing techniques. If not possible to deprocess within the package, then new techniques will be required to remove the die from its package before deprocessing.

REFERENCES

1 McGuire GE. *Semiconductor Materials and Process Technology Handbook.* New Jersey: Noyes Publication,1988.
2 Baerg W, Rao VRM, Livengood R. Selective Removal of Dielectrics from Integrated Circuits for Electron Beam Probing. Proceedings International Reliability Physics Symposium, 1992, 320.
3 Abramo MT, Roy E, LeCours SM. Reactive Ion Etching for Failure Analysis Application. Proceedings International Reliability Physics Symposium, 1992, 315.
4 Abott J. A Procedure for the Non Destructive Removal of Glassivation from Integrated Circuits. Proceedings International Symposium for Testing and Failure Analysis, 1986, 114.
5 Scacco P, Malberti P, Ciappa M. Wet-Etch of of Nitride Passivation Layers: An Effective Alternative to Plasma-Etch for Failure Analysis Proceedings International Symposium for Testing and Failure Analysis, 1994, 157.
6 Vanderlinde WE, Von Benken CJ, Davin CM, Crockett AR. Fast, Clean and Low Damage Deprocessing Using Inductively Coupled and RIE Plasmas. Proceedings International Symposium for Testing and Failure Analysis, 1996, 74.
7 Thornton JA. Magetron Sputtering: Basic Physics and Applications to Cylindrical Magnetrons. Journal Vacuum Science Technology,1978, (15), 171.
8 Campbell Stephen A., *"The Science of and Engineering of Microelectronic Fabrication,"* Oxford University Press, 1996, 273.

9

CROSS-SECTION ANALYSIS

Tim Haddock
Scott Boddicker
Texas Instruments Incorporated

While deprocessing is an excellent method for identifying many defects and relating them to electrical characteristics, a cross-section often provides more information about the defect itself. For example, the point in a wafer process at which a particle was introduced can be difficult to determine with deprocessing, but is usually obvious in a cross-section view. Cross-sections also provide information about certain types of defects that are inaccessible by deprocessing. For example, failures that occur at the bottom of etched holes such as vias and contacts are not readily observed after deprocessing. Issues that involve etch contours or the thicknesses of thin films are also not usually resolved by deprocessing. For many assembly-related defects internal to the package, a cross-section is the only way to directly observe a problem.

Just as smaller device features have required observation tools with better resolution (field emission scanning electron microscopes (FESEM), transmission electron microscopes (TEM), etc.), sectioning techniques have required higher precision. The newer techniques have not replaced the traditional ones; they have provided additional capabilities for those cases where the older techniques are inadequate. Table 9.1 shows the range of sectioning techniques used in the semiconductor industry, an estimate of their precision, and instances when they are typically used. This chapter covers each of these techniques, along with a discussion of staining, which is used in conjunction with some of the sectioning techniques.

9.1 PACKAGED DEVICE SECTIONING TECHNIQUES

When a packaging problem is analyzed, a cross section often provides important information. Typical cases where cross-sections are useful include lifted bonds, poor die attach, mold compound voiding, and package cracking. If sectioning of the die is desired, it is usually better to remove the die from the package and proceed with die polishing techniques.

For problems covering a significant area of the package, such as a large void, sawing the device may be all that is needed to view the defect directly. This is typically carried out using diamond bonded saw blades and is followed by whatever degree of polishing is required to make the large area defect readily observable. For most package problems, the package must be ground to the defect area as described below in Table 9.1.

Sectioning Technique	Precision (microns)	Typical F/A Uses
Packaged-unit sectioning: Sawing	2000	Gross internal package problems Ceramic Packages
Packaged-unit sectioning: Grinding	500	Assembly problems
Wafer cleaving: Manual	1000	Repeated structure or layer problems
Wafer cleaving: Precision	1.5	Localized structure or layer problems
Die polishing	0.2	Device defects > 0.5 microns
Focused ion beam: Sectioning	0.05	Surface defects or defects < 0.5 microns
TEM sample preparation techniques	0.05	Defects < 0.5 microns requiring high-resolution imaging

Table 9.1. Precision and uses of various sectioning techniques are shown.

9.1.1 Encapsulation

The general method of package sectioning is to encapsulate the package in epoxy (also known as "potting" material), and then to grind and polish to the desired location. The encapsulated package is easier to hold either manually or in a fixture for automated grinding and polishing. For very large packages, encapsulation may not be required to accomplish this. The encapsulation also provides a large area to grind and polish, making it easier to control the rate of material removal and to maintain planarity (see Figure 9.1).

These benefits of encapsulation are somewhat dependent on having a relatively uniform material hardness. This does not occur in the case of ceramic packages since ceramic materials are much harder. For ceramic packages, a diamond saw is used to cut as close as possible to the desired section plane, and then diamond discs are used to reach the area of interest. Encapsulation would be used for a ceramic package only if it was required to make the sample easier to hold or fixture.

For semiconductor applications, a clear two-part epoxy is preferred to other encapsulation materials. The device can be viewed through the clear epoxy with an optical microscope. This allows cross-sections of specific device features such as bonds. Various molds or mounting cups are used to determine the shape of the sample. Some can be used only once, while others are reusable and require a release agent with each use. The device can be oriented in the mold using a variety of methods, but a clip specifically designed for this purpose is the most common. After mixing, a mild vacuum is used to remove bubbles and moisture from the epoxy. A vacuum chamber vent is used to prevent overflows, associated with the outgassing. The epoxy is cured for a time that depends on the type of epoxy used and cure temperature. After curing and unmolding (if required), the sample is ready for grinding and polishing.

9.1.2 Grinding and Polishing

A wide range of abrasives is available. Both the suitability for a particular task and personal preferences are involved in the selection process. The three most popular abrasive materials are silicon carbide, aluminum oxide, and diamond.

Silicon carbide is a good general-purpose polishing material, often used for preliminary removal of large amounts of material. Aluminum oxide is also used for a variety of polishing tasks, as it is available in a wide range of particulate sizes. Diamond provides very good results for precision polishing, but it is considerably more expensive than the silicon carbide and aluminum oxide. Diamond is also the hardest of the materials. This can minimize the effects of hardness variations in the sample. Grinding and polishing samples containing materials of differing hardness leads to uneven material removal rates. The harder material is removed less rapidly compared to the surrounding materials. The edges of the harder material tend to round so that the interfaces are more difficult to observe. In some cases, the protrusion of the hard material makes it difficult to keep the section flat on the grinding or polishing wheel.

Figure 9.1. Polished cross-section of a packaged device is shown.

Each abrasive material is available in several forms, determined by ease of use and cost. All three materials are available with the abrasive bonded to a solid metal disc. They are also coated onto plastic sheets, referred to as films. Silicon carbide is commonly used in paper form. Aluminum oxide and diamond are also delivered as suspensions, which are applied to fabric discs. Low-nap versions of nylon and rayon are popular for fine polishing. Suspensions are available as liquids and sprays or can be prepared from powders. Table 9.2 summarizes the particulate sizes available for various forms of the common abrasive materials.

Another traditional and widely used polishing device is the glass wheel. This plate of hard lime glass has surface irregularities limited to much less than 1 micron, which produces a fine polished surface on typical semiconductor samples. Glass wheels last many months and require only periodic light detergent scrubbings to remove accumulated material. The glass wheel is most commonly used in the cross sectioning of dice.

Material	Form	<<Grinding>> <<Polishing>>
Silicon Carbide	Metal disc, paper	
	Film	
Aluminum Oxide	Metal disc	
	Film	
	Powder, suspension	
Diamond	Metal disc	
	Film	
	Compound, suspension, spray	
		1000 100 10 1 0.1 0.01
		Particle Size (microns)

Table 9.2. Particulate size availability for common grinding/polishing abrasives are summarized.

Grinding and polishing are performed on a rotating wheel designed for that purpose. The abrasive disc or cloth with suspension is usually held in place with a metal ring. The wheel normally turns at a variable rate from one hundred to a few hundred revolutions per minute. A small stream of water flowing over the wheel surface reduces sample heating from friction, removes material abraded from the sample, and also reduces the generation of airborne particulates. The sample can be held by hand or by fixture. Generally, the sample should be positioned so that the direction of polishing motion is from the harder material to the softer material at the interface of interest. This is particularly important for softer materials such as solder or gold. Soft metals can also smear, in which case a brief application of an appropriate acid can be used to dissolve to smeared metal. An added benefit is the decoration of grain boundaries by the acid. For example, aqua regia is an excellent decorating agent for gold ball bonds.

In typical applications, a series of particle sizes (also called grit sizes) from coarsest to finest is used to grind the sample to the desired location. Then a final polish using 0.3 micron or 0.05 micron alumina slurry on cloth is performed.

9.2 WAFER CLEAVING

The most fundamental wafer-level sectioning technique is manual cleaving of the wafer. This is possible because semiconductor wafers are processed with single crystal silicon usually having a <100> or <111> lattice plane surface. For CMOS, <100> is most commonly used, resulting in cleave planes along the wafer's x- and y-axes.

9.2.1 Manual Cleaving

Manual cleaving is both easy and quick. Once the desired area is located, the desired cleave line (x or y direction) is followed to the edge of the wafer and a scribe mark is made there using a diamond scribe in the same direction as the cleave line. A scribe mark about 5 mm in length with 1 or 2 firm strokes deep is sufficient. There are many methods for breaking the cleave. One of the most consistent methods is to place the end of a straightened metal paper clip under the scribe line along the direction of the desired cleave, followed by a slight downward pressure at the edges of the wafer a few centimeters away from the scribed line.

An accuracy of a few millimeters is readily obtainable with this method, and with practice 1-millimeter accuracy is routinely achievable. This method is most commonly used for repeated structures, such as an array of memory cells, when there is no need to section a particular cell. It is also useful for problems that are suspected to affect a thin film over a substantial area. Cleaving is also commonly used to get close to a point defect as a prelude to polishing, as described later. Cleaving is generally used on whole wafers or large portions of wafers, but it can be successful on pieces as small as a few cm^2.

Figure 9.2. SEM (Scanning Electron Microscope) image of a wafer cross-section prepared by cleaving is shown.

For sections where a key layer is a ductile material such as aluminum, better results can be obtained by immersing the sample in liquid nitrogen just before cleaving. The safe handling of wafers at liquid nitrogen temperature is somewhat cumbersome. A typical wafer cleave is shown in Figure 9.2.

9.2.2 Precision Cleaving

The accuracy of cleaves is greatly enhanced by semi-automated instruments developed specifically for cleaving. These systems integrate an optical microscope, a diamond-point scribe and a wafer tension/torque device. A coarse cleave positions the feature of interest near an edge. The defect is aligned to the microscope crosshairs. The instrument automatically cleaves the sample with an accuracy of 1.5 microns. It works best for a large piece of a wafer with a single target point. This method is very efficient for defects that are more than 2 microns in diameter, and for structures or thin film problems that are confined to a small area. A liquid nitrogen option provides low-temperature cleaving without exposing the operator to cryogenic hazards.

9.3 DIE POLISHING TECHNIQUES

For wafer features that require higher precision than cleaving can achieve, polishing is used. Polishing is also useful for devices that have been removed from the package. This technique is routinely used to section all types of defects or structures with dimensions greater than approximately 0.5 microns.

The sample, a die removed from a package or a small (1cm or less) piece of wafer, is mounted on a stub with the defect positioned beyond the edge of the stub. The mounting material is usually an epoxy or wax. The stub is attached to a polishing block, which has a Teflon edge. The arrangement is such that the stub and block combination rests on the Teflon edge on one side, and on the overhanging sample edge on the other side (see Figure 9.3). Next, the unit is placed against a fixture so that it rides on an abrasive disk on a rotating wheel. A moderate water flow is maintained over the wheel to reduce frictional heating. Periodic checks are made by removing the stub/block and viewing the device top surface with an optical microscope until the desired section location is reached. The stub remains fixed on block during inspections to maintain the parallel alignment of the sample edge with the Teflon edge.

Figure 9.3. Die mounting on a polishing fixture is shown.

Polishing can be ineffective for defects or layers that are at, or very close to, the surface. Surface features often break off during polishing. For this reason, a layer of protective oxide or nitride is often deposited over the surface defect to provide more mechanical stability to the area. This can be accomplished with spin-on glass or a plasma deposition system for a large area, or with a FIB (focused ion beam) system as discussed later for a point defect.

Ductile layers can also be problematic, especially when the layer of interest is at or near the surface. If there is a single interface of interest, polishing in a direction such that the smearing is away from the interface of interest can give satisfactory results. In addition, etching away the smeared material can decorate metal grain boundaries. In many cases, low-temperature cleaving or FIB sectioning may be more effective. See Figure 9.4 for an example of a polished cross section.

9.4 CROSS SECTION DECORATION: STAINING

Staining has become a general term applied to chemical delineation or decoration of device features, particularly in cross-section. Its main purpose is to provide contrast in imaging various device features. Cross sections have some natural contrast such as color in optical microscopy and atomic number and charging contrasts in SEM. Contrast can also arise from subtle differences in polishing rate for topographic techniques such as SEM. Improved resolution, as with FESEM, of natural contrast mechanisms can reduce the requirements for chemical decoration. Staining is required in cases where no significant natural contrast exists, for example, in viewing junctions in silicon with an optical microscope or SEM. Staining can also help delineate different layers of similar materials.

Figure 9.4. SEM image of a wafer cross-section prepared by polishing (unstained).

9.4.1 Common Stains and Uses

Many of the etchants used in deprocessing are also useful for cross-section staining of the same materials. Oxide etches typically involve HF (hydrofluoric acid) buffered with NH_4F (ammonium fluoride). Their etch rates are dependent on the density and doping of the oxide. "Common oxide etch" (COE) is a mixture of 6% HF (49% concentration by weight), 35% NH_4F (40wt%), and 59% water by volume. Another widely used oxide etch is known as "standard oxide etch" (SOE), which consists of 50% glacial acetic acid, 5% ammonium fluoride (40wt%), and 45% water by volume. This can be applied to the section for 20 seconds for a light delineation effect or up to 1 minute for a heavier delineation. Delineation can also be created in a plasma etcher or RIE (reactive ion etcher).

Junction decoration depends on variations in the etch rate with doping density and type. The more highly doped materials etch more quickly causing a relief in the doped areas (See figure 9.5). Lightly stained cross-sections usually provide more reliable junction depths. Junction staining can be accomplished with either copper-based electroplating or with silicon etches under certain conditions[1]. The active ingredient in mixtures for copper decoration is usually copper sulfate, which is mixed with hydrofluoric acid and nitric acid. A strong light source on the sample during application enhances the electroplating effect. Nitric-rich silicon etches also can stain junctions, the most common proportions being 200-1 and 50-1 ratios of nitric acid to HF. The junction stain can be immediately preceded by an oxide etch to clean the surface of any native oxide increasing the effectiveness of the stain.

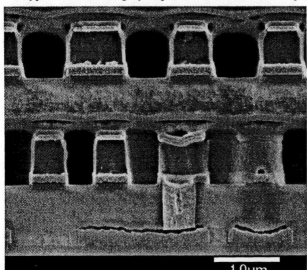

Figure 9.5. SEM image of a wafer cross-section prepared by polishing and stained with SOE.

9.5 FOCUSED ION BEAM (FIB) TECHNIQUES

The FIB has become an increasingly important tool due to its versatility, productivity and precision (also see Chapter 7). A major FIB application in wafer fab yield analysis and failure analysis is cross sectioning. A focused gallium ion beam rasters across a designated area, milling a rectangular hole into a sample. One side of the milled area becoming the cross-section plane. Several improvements in FIB technology have increased the usefulness and efficiency of FIB cross-section preparation. Dual column systems make it possible to mill with a FIB while simultaneously observing the sidewall of the cross-section with a FESEM. CAD navigation facilitates the use of device layouts to quickly locate an area to be sectioned. Wafer mapping capability allows rapid localization of defects identified by various in-line wafer fab tools. Improvements in ion beam resolution enable the sectioning of submicron features with precision and repeatability that is not achievable by mechanical sectioning techniques. The availability of gas injection to enable deposition and enhanced etching in FIB has increased the flexibility and efficiency of FIB cross sectioning.

A limitation to FIB sectioning is that the milling time scales linearly with the length of the section. For section lengths of 50 microns or more, other techniques tend to be more practical. A second issue is the beam damage to the sample. Redeposition and gallium implantation alter the surface being inspected. Charging from the ion beam has been observed to cause electrical overstress damage to some isolated structures.

9.5.1 FIB Sectioning with Enhanced Etches

FIB configurations allow gases to be injected through a needle near the sample surface. The gases are absorbed on the surface and undergo chemical reactions as a result of interaction with the ion beam. For organometallic compounds, the interaction with the ion beam results in decomposition depositing metal (usually Pt or W) on the surface. A metal strip is deposited over the planned section line to prevent any rounding effects at the top surface of the section. This is also useful for providing structural support for surface defects. A single-pass application of metal deposition on a FIB section lightly coats the section surface to reduce charging effects and enhance contrast.

Other gases can be injected which have the effect of selectively enhancing the milling of the ion beam. These include iodine, chlorine, bromine, and fluorine (xenon difluoride). These etch gases increase the etch rate, reducing the milling time. Since the etch enhancements are selective, brief enhanced etching can have a staining effect on FIB sections, analogous to staining of sections prepared mechanically. For example, the ion beam in a xenon difluoride enhancement will etch SiO_2 areas much faster than surrounding materials, delineating the oxide interfaces much as a HF-based stain would on a mechanical section.

9.5.2 Related FIB Uses

In addition to its precision and efficiency, the FIB has other sectioning advantages over standard techniques[2]. While surface defects or soft materials (like aluminum or photoresist) can break off or smear during conventional polishing, the FIB can cut cleanly through them. The FIB progressively sections through a feature, efficiently providing a succession of views. Both the ion beam and the electron beam in a dual column system can be conveniently used for final imaging, with each providing unique information due to the different contrast mechanisms for each (FIB cross sections from single column systems are commonly viewed in FESEM's as well). The FIB can also perform limited deprocessing, for example, opening up an area so that elemental analysis can be obtained.

FIB sectioning can be combined with cleaving or polishing. This is useful when the sectioning precision of the FIB is required, but the final imaging needs to be a true perpendicular view or at a resolution better than can be achieved in the FIB. Since the FIB section is basically a rectangular hole, the wall of the cross section must be viewed at an angle off orthogonal. First the sample is cleaved or polished to within a few microns of the point of interest. The FIB then mills out the remaining material to the desired location. Using a similar concept, the FIB can aid in the TEM sample preparation, as described later.

9.6 SECTIONING TECHNIQUES FOR TEM IMAGING

TEM is the technique of choice for very high resolution imaging of cross-sections. TEM images are used to identify many defect types that are not observable with a FESEM on standard cross-sections. The improvement occurs due to the increase in resolution of TEM versus the FESEM, as well as the different contrast mechanisms between them. Thin layers of contamination (for instance, in vias or contacts), gate oxide non-uniformity, and subtle etching problems are typical problems which require TEM. TEM is also a useful tool in the identification of silicon stacking faults and dislocations. TEM imaging can also be used to identify chemical phases since lattice spacings of crystalline materials can be resolved. Lattice spacings are also used for magnification calibration, which permits very accurate measurements of layer thicknesses.

The drawback of TEM is the size limitation of the sample. The sample must fit on a grid that is 3 mm in diameter. In addition, the area of interest must be thinned to be transparent to electrons. The thickness requirement is often stated to be less than 100 nm, but for best results the sample thickness should be closer to 50 nm. The majority of the effort for TEM viewing of semiconductor samples is preparing the sample to meet these dimension requirements. Fortunately, improvements in the efficiency of sample preparation for TEM have transformed it from a research technique to an important defect analysis and characterization tool.

9.6.1 Stacked-Sample Technique.

One common technique for preparing TEM cross-section samples is known as "multiple stacking". This technique is useful in cases where a precise section line is not required, such as large arrays or general layer viewing. It allows for preparation of multiple sites at one time. The process is summarized is Figure 9.6.

The areas of interest are cut by diamond saw into pieces that are 3 mm by 1.5 mm. If there are less than four samples with areas of interest, additional dummy pieces are sawed to bring the total up to four. Each piece may be used to inspect a different area. The pieces are glued into a stack using an appropriate (low viscosity and minimal outgassing, such a phenolic-epoxy adhesive) adhesive. The total thickness of all the pieces is limited to 2 mm.

There can be more than four pieces in a stack as long as the total thickness fits within the viewable area of a TEM grid. Traditionally, this has limited the number of pieces to six, but backside grinding of the pieces has allowed for many more samples to be processed in the same stack[3]. The "fronts" of the pieces are oriented in a non-symmetric pattern to permit unambiguous identification of each piece when imaged by TEM.

After curing the glued stack in a vise, the thinning process begins. The stack is attached with a thermal wax to a polishing stub (the polishing tool described later for the wedge technique can be used). The sample is polished on a polishing wheel under flowing water using successively finer diamond films, from 30 microns to 0.5 microns, until about half of the stack width is removed. A final polish with a 0.05 colloidal silica suspension on a low nap cloth completes the sequence. The stack is removed by heating the wax, and the polished side is glued (again with a TEM-appropriate adhesive) to a TEM grid. The grid/stack sample is attached with thermal wax to the polishing stub and the other side is polished, this time to a remaining thickness of about 30 microns. The sample is removed and rinsed with a solvent to remove any wax residue.

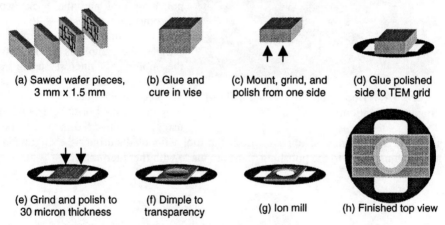

Figure 9.6. Steps in preparing a TEM sample by the stack method.

A dimpler is used to thin out the center of the sample to light transparency. This is achieved with a sequence of diamond pastes from 3 micron to 0.5 micron. Then an argon ion mill continues the low-angle thinning about the sample center, until a hole is formed that intersects all of the pieces of the sample. TEM images are taken from around the edge of the hole where the sample is extremely thin.

9.6.2 Wedge Polishing Technique

For cases where the feature of interest is at a specific point, the "wedge" technique was developed[4]. This technique uses a special tripod polishing tool (often referred to as a "Klepeis polishing tool" after its developer) to polish the sample to a wedge shape, with imaging done at the tip of the wedge (see Figure 9.7 and 9.8). Since this is done manually, a high degree of concentration and tactile skill is required for success.

170   Cross Section Analysis

Figure 9.7. Tripod polishing tool for TEM sample preparation.

Initially, the sample is cut by diamond saw into a piece about 3 mm square with the point of interest in the approximate center. The sample is wax-mounted onto the side of the glass polishing stub insert with the top sample surface facing away from the tripod, and with the point of interest slightly overhanging the stub edge.

With the stub in place on the polishing tool and the back two micrometers adjusted to be coplanar with the bottom of the sample, the sample is polished to the point of interest using a sequence of diamond films as in the stack technique. Viewing the progress of the polishing is easiest using an inverted microscope, since the stub cannot be removed from the tool without disturbing the alignment. Minor adjustments to the polishing plane are made with the micrometers.

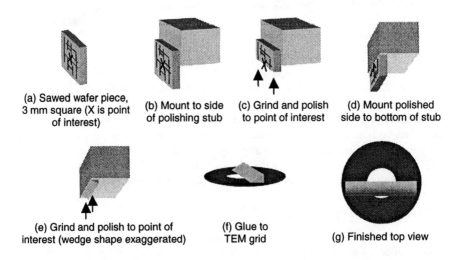

Figure 9.8. Steps in preparing a TEM sample by the wedge method.

The sample is unmounted and cleaned, and then the polished sample face is wax-mounted to the bottom (rather than the side) of the stub insert. Again, the top of the original sample should be facing away from the tripod. The back two micrometers are aligned with the sample bottom, and then they are advanced equally about 1 mm. This produces a slight wedge shape as the polishing sequence proceeds as before. Polishing is complete when the apex of the wedge (at the point of interest) become optically transparent as indicated by clear diffraction fringes.

The sample is then removed from the stub by soaking over filter paper, and a TEM grid is carefully glued onto the sample. Some minor ion milling may be necessary before TEM imaging.

9.6.3 FIB-Prepared Samples

FIB is increasingly used for TEM sample preparation[5]. Like the wedge technique it is used for TEM at a specific point, but the FIB system reduces the reliance on

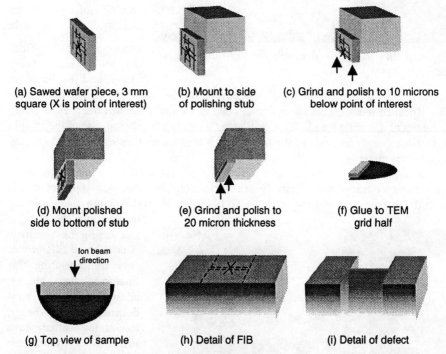

Figure 9.9. Steps in preparing a TEM sample using a FIB

manual skills. In essence, two FIB cross sections approach the point of interest from either sides, leaving a thin membrane. This process (See Figure 9.9) significantly reduces the time required for TEM of a specific feature. On the negative side, it is not possible to eliminate implant damage from the gallium beam, which locally disrupts crystallinity.

The bulk sample preparation is similar to that for wedge samples. A grid-sized sample is polished with successively finer grained diamond film. The difference in this case is that the polishing of each side is stopped when the point of interest is about 10 microns away. This leaves the target in the approximate center of a 20 micron thick sample. The sample is glued onto a TEM slot grid that has been cut in half. Removing half of the grid keeps it from interfering with the FIB beam.

Special vises are available that hold a TEM grid vertically for FIB milling. A light coating of gold can eliminate possible charging problems from the ion beam. Once inside the FIB, the target line is marked and a strip of metal is deposited on the surface along the target line. The deposited metal serves as sacrificial material to minimize the effects of rounding at the top of the milled wall. Bulk-mill areas are set up from each side of the sample to within a couple microns of the target, and a high-current beam is used to quickly remove this material.

Figure 9.10. FIB-prepared sample for TEM imaging is shown.

Starting from one side, standard FIB techniques are used to section to the target line. The sectioning proceeds in the same way from the other side. Greater care is needed for the final milling to leave a film that is 500-700 nm thick without overmilling the sample and losing its integrity. A dual-beam FIB proves very useful during this final foil thinning since it allows for non-destructive (e-beam) monitoring after each FIB slice. An example of a final foil is shown in Figure 9.10 and a typical TEM from a FIB section is shown in Figure 9.11.

Figure 9.11. ATEM image from a sample prepared by FIB is shown.

Once the proper foil thickness is achieved, the sample is ready for TEM imaging. There is still some delicacy involved in polishing the sample to about 20 microns thickness and mounting on a grid, but the most critical part is now a more repeatable process.

Some work has been done to try to eliminate mechanical polishing altogether[6]. The FIB sectioning proceeds as above on a large bulk sample. The foil is then tilted with respect to the ion beam, and FIB cuts are made through the two sides and the bottom of the foil. Only a small section on one side of the foil is not milled through so that the foil remains attached to the bulk sample. The sample is removed from the FIB and a micromanipulator is used to electrostatically detach the foil from the bulk material. The detached foil is then placed on a formvar-coated TEM grid. The formvar coating is necessary since the size of the foil is much smaller than any available TEM grid spacing. The formvar both supports and adheres to the foil. This technique eliminates the skill involved in polishing down to 20-micron thickness, but requires the additional skill of manipulating a microscopic sample onto a TEM grid.

9.7 FUTURE ISSUES

Changes in sectioning techniques will continue to be driven by the constant shrinking of features of interest. The primary effect will be to push more samples into the FIB and TEM realms. The best case FIB precision is about 5 nm at present, which perhaps can be improved by a factor of two over the next decade. The main issue with FIB is the damage from the ion beam. This should not be a significant problem for standard surface imaging, but it could become a limiting factor for TEM imaging of certain samples. As feature sizes shrink and require thinner TEM samples, the damage layer from FIB preparation will extend through the width of the sample. This will have the most impact where crystallinity or composition issues are important. There is no obvious solution for this problem, but it may be mitigated by the development of FIB sources other than gallium.

REFERENCES

1 Lee TW. A Review of Wet Etch Formulas for Silicon Semiconductor Failure Analysis. International Sympsoium for Testing and Failure Analsyis, 1996, p. 319.

2 Haddock T, Smith B, Tsung L. Failure Analysis Techniques for Wafer Fabs. Texas Instruments Technical Journal, Oct. 1997, p. 24.

3 Basham PB, Tsai L. Advanced TEM Sample Preparation Techniques for Submicron Si Devices. Proc. Microscopy and Microanalysis 1995, Jones and Begell Publ., 1995, p. 516.

4 Benedict J, Anderson R, Klepeis S.J. Recent Developments in the Use of Tripod Polisher for TEM Specimen Preparation. Proc. Materials Research Society Symposium Proc., 1992, vol. 254, p. 121.

5 Basile DP, Boylan R, Baker B, Hayes K, Soza D. Focused Ion Beam Milling for TEM Sample Preparation. Proc. Materials Research Society Symposium Proc., 1992, vol. 254, p. 23.

6 Overwijk MHF, van den Heuval FC, Bulle-Lieuwma CWT. Novel Scheme for the Preparation of Transmission Electron Microscope Specimens with a Focused Ion Beam. J. Vac. Sci. Techn. B, vol. 11, 1993, p. 6.

10

INSPECTION TECHNIQUES

Lawrence C. Wagner
Texas Instruments Incorporated

The process of determining the physical cause of an IC failure is typically a combination of sample preparation and inspection. The most common sample preparation techniques, deprocessing and sectioning, have been discussed in Chapters 8 and 9. Deprocessing and inspection form an intertwined process of finding and identifying defects. In most cases, it is a truly interactive process with some sample preparation followed by inspection leading to additional sample preparation and inspection until the defect has been exposed and identified. The inspection process for IC failures takes advantage of a number of forms of microscopy. Optical microscopy has historically played a key role in integrated circuit inspection due to its ease of use and interpretation. With shrinking feature and defect sizes came the demand for higher resolution, which has been provided by the SEM (Scanning Electron Microscope). Ultimately, even SEM resolution has proven inadequate for some applications leading to the expanded use of TEM (Transmission Electron Microscope) and SPM (Scanning Probe Microscope).

10.1 MICROSCOPY

Typically, there are several key parameters to be considered in selecting from the various microscope forms available. Since IC features and defects have become extremely small, resolution is perhaps the important factor in selecting microscope types. The importance of resolution in IC microscopy is expected to further increase as feature sizes continue to shrink. Resolution is usually defined in terms of the smallest gap between objects, which can be resolved by the microscope. Thus, a numerically smaller resolution is better. While there are specific features of microscopes which impact resolution, some general features are common. The laws of physics limit resolution of a microscope to something on the order of magnitude of the wavelength of the incident radiation. This is commonly referred to as the diffraction limit. Thus, the wavelength of the incident radiation can be used to impact the maximum resolution theoretically obtainable. For example, in optical microscopy the diffraction limit is defined by a constant (typically 0.5-0.6 depending on the optical system and definition of resolution) times the wavelength of light divided by the numerical aperture. The diffraction limit is apparent in the differences between IR and visible light microscopy. IR light microscopy images typically exhibit degraded resolution compared to a visible light microscopy due the longer wavelength of light and the diffraction limit. Conversely, UV light microscopy should provide an improvement in diffraction limited resolution. The most visible example of this occurs in photolithography where improvements have consistently been made by using shorter wavelengths of light. This is also reflected in the use of monochromatic light. It is better to use a short wavelength from a band rather than a full band of light, since the diffraction limited resolution will be improved. In addition, the use of monochromatic light eliminates any chromic aberration issues. Chromic aberration is variation with wavelength in the focus of radiation through lenses. Radiation of different wavelengths will tend to focus in a slightly different plane. This applies equally well to charged particle beams where a distribution of electron or ion wavelengths (or energy) can result in significant chromic aberration.

The source of radiation is also a significant factor in the resolution of techniques. Resolution is generally enhanced by making the source of radiation appear more point-like. The diagram in figure 10.1 shows the impact of source size on resolution. This very simple example is the real case of X-radiography. A simple method of reducing the effective source size is the addition of an aperture as shown in figure 10.1. More complex situations with lens and mirrors lead to similar results. In many forms of microscopy, apertures and lens provide methods of improving resolution.

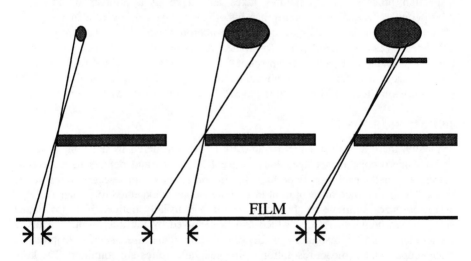

Figure 10.1. Resolution can depend on the size of the radiation source. The effect of source size is shown in a simple example of an edge exposed to radiation from a large and small source. The "gray" area or resolution of the edge is larger for the large source. Adding an aperture to larger source as shown on the right improves the resolution.

For scanning microscopes, the source radiation is scanned as a spot across the sample in a raster. The image is generated by mapping the intensity of the resulting radiation (e.g. reflected light for scanning laser microscope or secondary electron emission for a SEM) to the physical location of the beam spot. Resolution in scanning microscopes is highly dependent on creating a small spot from the source radiation. Creating the smallest possible spot usually requires some sacrifice of some signal intensity. Thus resolution is commonly traded off for contrast which is a reflection of the signal to noise ratio. In scanning microscopy, the function of the lenses and apertures is to provide a very small spot at the focal point of the lens system. The importance of this can readily be seen by taking the simple example of a spot which is large relative to the object and scanning the spot across it as shown in figure 10.2. Another factor in scanning microscopy is spot shape. Deviations from a circular spot are called astigmatism. As shown in figure 10.2, when a distorted spot is projected the image is distorted. Astigmatism can also be a problem in non-scanning microscopes where it creates distortions in the image.

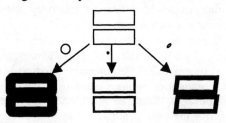

Figure 10.2. The resulting image from scanning various spots across a pair of objects is illustrated. From left to right, a large spot, a small spot and astigmatic spot

Depth of field is also a significant factor. Depth of field is basically the range of heights over which a microscope image is in focus. Depth of field impacts the ease of use of microscope and can also impact the viewing of features, which are not in the same plane. Generally, the larger the depth of field, the easier a microscope is to use. From a physics perspective, the factors, which impact resolution, also impact depth of field in the same direction. Thus as resolution improves, depth of field often becomes smaller or worse.

A third factor is simply how the microscope operates and what is being observed. Some microscopes such as the SEM provide a topological view, while other microscopy can take advantage of the transparency of various materials to observe buried objects such as optical microscopes.

The dominant forms of microscopy for IC inspection, have been optical microscopy and scanning electron microscopy. As resolution requirements become greater, TEM and SPM in its various forms are becoming more important. In addition, with the diverse applications of the Focused Ion Beam becoming important, an understanding of ion beam imaging is important.

10.2 OPTICAL MICROSCOPY

Optical microscopy has a number of advantages. One advantage is its ease of use. No vacuum is required making sample loading simple. The images are very intuitive and easily interpreted. In addition to intensity variations, color plays a significant part of the contrast and observability of defects with optical microscopy. Another advantage is versatility. Optical microscopy provides several unique modes of operation, which enhance its effectiveness as a failure analysis tool. In terms of its interaction with semiconductor devices, perhaps the most significant advantage is the transparency of most of the thin film dielectrics used. This allows viewing through these dielectric layers. This advantage is unique to optical microscopy because higher resolution techniques tend to be primarily topographical in nature.

The disadvantages of optical microscopy, its limited resolution and depth of field at high magnifications, have become more significant as feature sizes have decreased and the number of metallization layers has increased. State-of-the-art feature sizes have fallen below the diffraction limit of visible light. Although, objects, which have sizes below the diffraction limit, can be detected as distortions, they can not be resolved into a useful image. The advantages of optical microscopy are so significant that efforts have been made to extend the useful life of optical microscopy. As a result, confocal and laser scanning microscopes have become relatively standard inspection tools since they optimize optical microscope resolution.

Optical microscope resolution is limited by physics to approximately half the wavelength of the light used. In addition to the linear relationship to wavelength, the resolution is inversely proportional to a quantity called the numerical aperture or NA. NA can be viewed as a measure of the amount of light collected from the sample. There is a numerical aperture for the optical platform and a numerical aperture for the objective lens, which are averaged for the overall numerical aperture. For air lenses, physics limits the NA of lens to a maximum value of 1. For immersion lenses, NA apertures above 1 are achievable with corresponding improvements in resolution. Air lenses are limited by the refractive index of air, while immersion lenses allows the use of higher index of refraction materials. Immersion lens employ a liquid material, typically oil or water which have a higher index of refraction than air. A drop of the material is used on the bottom of the objective lens and is in contact with the device under observation.

In a standard optical microscope, the purpose of the objective lens is form create a focal plane image of the device being inspected. Placing a camera or detector in the focal plane can capture the image in this focal plane. Alternatively, this image can be recreated with ocular lens to be focused on the retina for viewing. Electronic imaging has greatly reduced the use of film for optical micrographs. Electronic imaging facilitates the generation of fully electronic failure analysis reports. Film is now used in only the most demanding situations such as feature size measurements.

Depth of field is predominantly influenced by the same factors as resolution. Thus as resolution improves (decreases), depth of field becomes smaller. Generally as resolution is improved, depth of field is degraded. Confocal microscopy provides two techniques to overcome this limitation by integrating images of differing focal planes.

Working distance is also a significant number for optical microscopy applications in failure analysis. In many failure analysis applications, long working distance objective lenses are required for inspection of devices in a package. High numerical aperture lenses have very short working distances. An objective lens farther away from the sample will be able to collect less light from the sample, giving it the lower NA. The short working distance lenses are not a serious issue for imaging wafers. However, such short working distance lenses can limit inspection of packaged devices. In general, packaged device failure analysis takes advantage of longer working distance lenses, which may have slightly lower numerical aperture. Lenses for mechanical probing tend to be ultra-long working distance to accommodate probes, yielding the poorest resolution but best depth of field.

10.2.1 Brightfield

Brightfield is the most frequently used type of optical microscopy. In brightfield illumination, the incident light is sent through the objective lens using the same path as the reflected light which will collected in the eyepieces or camera (figure 10.3) This illumination method for optical microscopy is particularly useful for flat samples. Since semiconductors are extremely flat, brightfield has become the dominant application. Brightfield microscopy, however tends to provide rather poor visibility of rough surfaces. Light incident on a rough surface is reflected over a broad range of angles (see figure 10.4). As a result, very little of it is reflected back into the objective lens. This causes any rough areas or bumps to appear dark in brightfield.

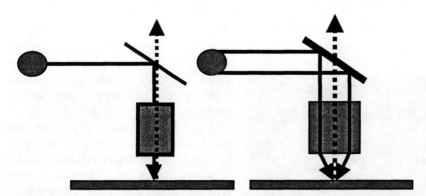

Figure 10.3. The path of light through an optical microscope is shown. For brightfield illumination (left), the path the incident light (solid) is coincident with the reflected light. For dark field illumination, the incident light (right) is oblique on the sample.

10.2.2 Darkfield

Darkfield microscopy provides an alternative illumination path for the incident light beam as shown in figure 10.3. The incident light is directed through the outside of the lens barrel making the illumination oblique. Very flat samples appear dark in

180 Inspection Techniques

darkfield because the oblique lighting is not reflected back into the objective lens. While darkfield illumination is less commonly used in semiconductor applications, it can be a useful tool for inspection of objects that appear dark in brightfield.

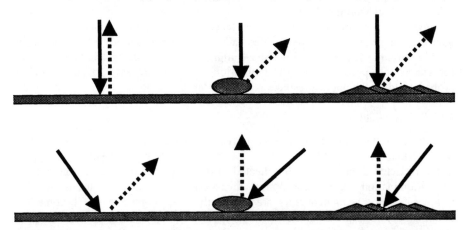

Figure 10.4. Reflections from various surfaces are illustrated. In brightfield illumination (top), only light from the flat surface is reflected back into the objective lens. For darkfield (bottom), only light reflected from rough surfaces is reflected into the objective lens.

10.2.3 Polarization and Interference Contrast

Creating higher contrast in optical microscope images can provide improved viewing of images. The addition of polarizing lenses functions much as polarizing sunglasses to reduce glare or spurious light reflections. For example, the use of polarized light in liquid crystal analysis greatly increases the contrast at hot spots. By adding a diffraction element called a Wollaston prism to the optical path, variations in the height of a sample can be accentuated. The Wollaston prism splits the light into two orthogonally polarized beams. In many cases, the resulting interference between the two beams is more sensitive to subtle changes in contour than SEM. Interference contrast remains the primary method of observing silicon defects as illustrated in figure 10.5. However, interference contrast can also be

Figure 10.5. Interference contrast image of bond pad after deprocessing and decoration for defects.

a useful tool in cross section inspection to delineate poorly defined features and other applications.

10.2.4 Confocal and Laser Microscopy

Confocal microscopy is used as a method of increasing resolution for optical microscopes. As pointed above, one technique for improving resolution is making the source of incident radiation appear to be coming from a single point. This can be achieved by actually reducing the size of the source or by using apertures to reduce the apparent size of the source as illustrated in Figure 10.1. In confocal microscopy, the aperture is reduced to a pinhole, creating the maximum resolution. Figure 10.6 illustrates the effect of confocal microscopy. There are several trade-offs made to overcome the improved resolution. The first is that the amount of light transmitted through a pinhole is very small. The second is that, depth of field is sacrificed. There are two distinct approaches to overcoming these issues.

One approach to confocal microscopy is to use a large number of pinholes at the same time. These pinholes are located on a spinning disc. Multiple pinholes provide greater light throughput. Different pinholes can also provide different focal planes and provide a virtual depth of field. This also has the advantage of providing a real time image. This approach provides a real time image with optimized resolution and depth of field.

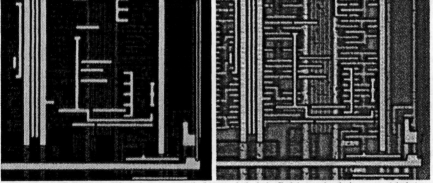

Figure 10.6. A confocal image (left) and brightfield optical image (right) are compared. Note that the metallization lines are more crisped defined in the confocal image but depth of field is degraded.

The second approach is to use a laser. This provides a more intense source of light through a single pinhole. Typically laser confocal microscopes are scanning systems where the laser is mechanically scanned across the surface. While this approach overcomes the low light level issue, the issue of depth of field remains. Laser scanning confocal microscopes (LSM) typically generate depth of field by summing multiple images with different focal planes. Images are collected at multiple planes (various heights) and added. Since out-of-focus areas in confocal microscopy are black, no correction is required for out of focus areas in any of the images. Since a significant number of images must be acquired in a scanning mode, real time images are not practical. Scanning laser microscopes also find applications in a number of failure site isolation techniques (see Chapter 6).

10.2.5 Ultraviolet and Infrared Microscopy

While visible light microscopy has the advantage of being a very natural method to observe devices, light with wavelengths outside of the visible region (approximately 400-800nm) has some significant advantages for microscopy. Both longer and shorter wavelengths of light are used in different applications. While these techniques present opportunities to do things not possible in the visible region, they are also are significantly different in some ways. Since the light outside the visible region can not be directly observed through an eyepiece, a detector of some type must be used to collect the image. Like visible light microscopy, the image is focused to a plane where the detector is located. Most detectors collect an image pixel by pixel. Unless the pixels can be updated at a television rate or faster, the images will tend to "jump" as the object being observed is moved. This makes navigation much more difficult. This is particularly a problem for observing highly repetitive structures such as memory arrays where counting cells can be made impractical without smooth image movement. A second disadvantage is that color does not retain its natural meaning. It is possible by get faux color through wavelength discrimination. While this can help to enhance contrast, it does not retain its intuitive meaning.

Infrared microscopy has been used for many years[1] to inspect device features through semiconductors. Silicon and other semiconductors exhibit transparency in the infrared region which makes it possible to "look through" the semiconductor material at the bottom of device features. The disadvantages of IR microscopy include several inherent factors. The IR wavelengths are longer than visible light. Thus, the diffraction-limited resolution of IR microscopy is not as good as visible light microscopy, which is itself becoming marginal for state-of-the-art devices. Although the resolution is very limited, the combination of migration to flip-chip packaging and increases in the number of interconnect layers make backside viewing through silicon a highly desirable feature of IR microscopy. It is now used as the method of navigation for virtually all of the backside failure isolation techniques (chapters 5-7). For wavelengths of light slightly above the bandgap of silicon, IR light can be used as an active probe, i.e. the light has enough energy to generate electron-hole pair formation. At longer wavelengths (less energy), IR light does not have enough energy to create carriers and can provide an electrically passive probe. However just as dielectric transparency is a major driving force for visible light microscopy, semiconductor transparency is a prime motivator for using IR microscopy. IR microscopy has some drawbacks.

Improvements in diffraction limited resolution require the use of shorter wavelengths of light. The nearest band of light with shorter wavelengths is the ultraviolet (UV). This is also the direction in which photolithography has migrated in order to improve resolution. In an effort to increase resolution, it is desirable to use monochromatic light rather than a band. Since color does not have a natural meaning, monochromatic imaging imposes little penalty while eliminating chromic aberration. Like IR microscopy, it must employ a conversion system in order to view an image just as IR light. A key factor is that through at least a part of the UV spectrum, silicon oxide and silicon nitride remain transparent, extending a key advantage of visible light microscopy.

For both IR microscopy and UV microscopy, many of the same techniques used with visible light can be used to improve the resolution. The use of lasers allows reduction in apertures while maintaining a reasonable signal level. This also includes confocal microscopy. In addition, the numerical aperture of the systems can be further improved using immersion lenses with high refractive index media.

10.3 SCANNING ELECTRON MICROSCOPY

Scanning electron microscopy[2] (SEM) provides an improved level of resolution compared to optical microscopy. As described above, resolution in scanning microscopes is dependent on creating a small spot, which can be readily achieved with apertures. In addition to forming a small spot, the usefulness of the image is dependent on achieving a useful signal to noise ratio. Since the output signal is dependent on the input radiation, a small but very intense spot is required. Early SEM's used a heated hairpin tungsten filament. Beams with a factor of 2 higher intensity were achieved with lanthanum hexaboride (LaB_6) rods. However, a much greater improvement was achieved with field emission electron sources. The FESEM (Field Emission Scanning Electron Microscope) provides useful magnification to several hundred thousand times. In fact, field emission sources have become the standard for semiconductor electron beam tools ranging from critical dimension measurement tools to e-beam probers to Auger Spectrometers etc.

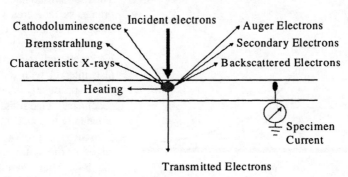

Figure 10.7. Processes resulting from electron irradiation are summarized.

Electron irradiation of a sample generates a broad range of emissions as shown in figure 10.7. In SEM imaging, primarily secondary and backscattered electrons are detected. In addition to these emissions, carriers can be generated in the device by incident electrons, providing the basis for several failure site isolation techniques, discussed in Chapters 6 and 7. Chemical analysis can be performed using characteristic X-rays generated or Auger electrons as discussed in chapters 12 and 13. The sum of these processes is not charge-neutral for the sample resulting in a net charge generated on the sample. It is important to understand that the created radiation can also have a high energy and generate its own stream of events. For example, backscattered electrons exiting the device can generate secondary electrons at the surface as they exit. This is critical because it can cause secondary electrons to be created in an area outside of the primary beam spot, effectively reducing resolution.

Secondary and backscattered electrons are very different in energy. Since secondary electrons (Figure 10.8) have very low energies, they can be actively collected. They can be attracted by a positive potential on a grid in front of an electron detector. On the other hand the higher energy backscattered electrons are not significantly impacted by the small potential on the grid. In fact, by placing a small negative potential on the grid, second electrons can be prevented from reaching the detector, making it a backscattered electron detector. More typically, a dedicated backscattered electron detector is employed.

10.3.1 Contrast Mechanisms

Contrast in SEM images arises from several sources. The primary contrast mechanism is topological. Contrast arises from topography due to expansion or contraction of the area for surface interaction with the electron spot and hence increasing or decreasing the secondary electron emission. Thus an irradiated spot of greater surface area will emit more secondary electrons. Exiting X-rays and backscattered electrons can also interact with the surface to produce additional secondary electrons further enhancing this effect. This is particularly pronounced at edges in the sample where a much greater part of the electron interaction volume becomes exposed on the sidewall. These effects also

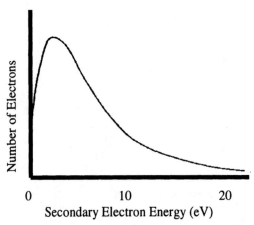

Figure 10.8. Secondary electrons exhibit very low energies, predominantly in the 0-20 eV range.

account for the increase in secondary electron intensity with sample tilt angle (see Figure 12.3). In addition to increasing the secondary electron signal, increasing the tilt angle provides a three dimensional perspective. The image observed from a tilted sample appears as viewed from the electron beam source rather than the detector. Since contrast in SEM is primarily topological, transparency of materials can not be used to observe buried features. Buried features are only observable to the extent that they alter the surface topography or impact one of the other contrast mechanisms such as charging.

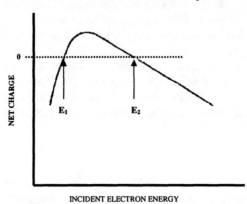

Figure 10.9. The net surface charge in a SEM is a function of electron beam energy. Charge neutrality is achieved at E_1 and E_2.

The second important contrast mechanism, particularly in IC's, is charging. As pointed out above, the net effect of the all of the processes generated by the electron interactions is usually not neutral in terms of charge. The charge on the surface results in a local electric field. Since the secondary electron energies are very low, this local field will have a significant impact on how readily the secondary electrons escape from the surface. However, there are accelerating potentials where charge neutrality occurs, commonly called cross over voltages, E_1 and E_2, as shown in Figure 10.9. In practice, excessive charging will make imaging unstable and can ultimately impact the primary electron beam as well as the secondary electron emission. SEM examination of IC's often results in a broad range of charging behavior across a sample of interest. Metal interconnects may be well grounded while the dielectrics often acquire a charge.

Atomic number also contributes to secondary electron contrast but the impact is small compared to the atomic number impact on backscattered electrons.

10.3.2 Backscattered electrons

Backscattered electrons provide an alternative method of imaging. In essence, backscattered electrons are incident electrons, which have bounced off the nuclei in the sample. The primary contrast mechanism in backscattered electrons is atomic number. This contrast arises because electrons are much more likely to be backscattered from a large nucleus than a small nucleus. Backscattered electrons are typically much higher in energy that secondary electron. The center of energy distribution for backscattered electrons is typically 50-95% of the incident electron energy. Light elements tend have the lower energy distributions along with lower intensity. Backscattered electrons can be readily detected with the standard SEM electron detector with secondary electrons rejected by negatively biased grid. As long as there is line of sight between the detector and sample, backscattered electrons will contribute to SEM images, providing much of the atomic number contrast.

More commonly, backscattered electron imaging uses segmented solid state detectors, which are mounted at the top of the SEM chamber and samples are inspected flat. Since backscattered electrons are strongly directional, not significantly deflected by the local electrical fields, backscattered images depend on the direction of detector from the sample. Backscattered images are particularly useful in analyses involving heavy metals. For example, detection of lead rich areas of solder is relatively easy with a backscattered image.

10.3.3 SEM Variables

There are actually a limited number of variables in SEM work. One of the key variables is the incident electron energy or accelerating potential of the beam. Resolution in SEM's is diffraction limited just as light microscopy. Since electrons exhibit particle-wave behavior, they have a wavelength, which results in a diffraction-limited resolution. The wavelength of the electrons is inversely proportion to the energy of the electrons (or the accelerating potential) of the SEM. However, the advantages of higher accelerating potential for resolution are often offset by other factors. Higher accelerating voltages result in more energy being available for creating secondary radiation. Since secondary radiation can result in secondary electrons on exiting the sample, higher energy beams can lead to more secondary electron emission from outside the incident electron spot. This, in fact, degrades the spatial resolution. Higher accelerating potentials also tend to create greater sample charging as previously observed. Available beam energies are typically in the 1-25 KeV range. Typically, beam energies of 5-25 KeV are most commonly used for samples where charging is not an issue. Lower beam energies (1-5 KeV) are used for samples that charge.

Beam current is also significant variable in electron beam applications. Reducing the beam current allows the creation of smaller spot with a higher resolution. At the same time, the reduced beam current degrades the signal to noise ratio. This has a major impact on contrast and image quality. In electron beam instruments, which observe lower cross-section event such as Auger, higher beam currents are a requirement for an acceptable signal. Poor signal to noise results in a grainy image that can be improved by averaging several images. However, navigation on a sample requires a real time image (television rate scanning), particularly in repetitive structures. Generally, beam current is set to the minimum value that generates an acceptable signal to noise ratio.

Working distance in the SEM has a significant impact on resolution. Ideally, SEM work should be performed at a very short working distance. The electron optics or lenses serve to focus the electron beam to a spot. As the working distance increases, the distance from the optics increases and the beam becomes more difficult to control and keep focused. In fact, the highest resolution SEM's place the sample inside the final electron lens. These "in-the-lens" SEM's are limited by a small sample size requirement. One problem created by using very short work distances is collection of the secondary electrons. As the working distance is shortened, the path from the sample to the secondary electron detector tends to be cut off. For very short working distance work, detectors located in the final lens provide a way to maximize signal collection.

Increased tilt angle has several benefits. As discussed above, the secondary electron intensity is increased with increased tilt angle. This provides better signal to noise and the three dimensional perspective of SEM's. On the hand, for large samples, increasing the sample tilt can result in very long working distances.

10.3.4 SEM Depth of Field

In addition to exhibiting excellent spatial resolution, the SEM exhibits very useful depth of field at lower magnification. In addition to providing higher resolution than optical microscopes, the SEM provides better depth of field. This is particularly important in package evaluations. For example, many evaluations of bonding require the depth of field possible with SEM as illustrated in Figure 10.10.

10.3.5 Sample Preparation

The preparation of samples for introduction into the SEM sample chamber is generally centered on reducing charging effects. Grounding the sample is a key factor in obtaining high quality images. Grounding provides a mechanism for bleeding off any accumulated charge. A path from the sample surface to the ground must be provided. The sample holder is typically grounded through SEM hardware. The sample can be attached and electrically grounded to the holder in several ways. In general, mechanical attachment such as clips should be preferred to the use of conductive paints (typically silver or carbon) since the paints can degrade the sample chamber vacuum.

Establishing a path from the device surface through the sample is the more difficult issue. Establishing this path is made difficult by the dielectrics in the IC, which can prevent discharge. This is particularly true for deprocessed devices where the electrical connections through bond wires and metal interconnects are severed. The preferred alternative is to use

Figure 10.10. The SEM image of a bond wire touching a stitch bond post illustrates the good depth of field at low magnification.

188 Inspection Techniques

low beam energies, typically around 1 KeV. When this is not possible or not effective in reducing charging effects, sample coating can be used. Very thin films (a few monolayers) of a conductive material are typically sputtered onto the sample. Chromium is the preferred coating material for FESEM applications. Gold or gold alloys were commonly used prior to the use of FESEM. These materials deposit in islands, which are not resolved by SEM's with thermionic emission sources (W and LaB_6), but are resolved by the FESEM's. Carbon coating can also be used when EDS analysis is to be performed. In addition the issue of artifact introduction, coatings can also be difficult to remove for further deprocessing. Gold films can be removed in iodine/potassium iodide solutions but chromium and carbon are not readily removed. Long SEM exposures can also make further deprocessing difficult due to surface contamination. The impact of this contamination can be reduced by good vacuum practices and minimizing unnecessary SEM rastering in the areas of interest.

10.4 FOCUSED ION BEAM IMAGING

Focused Ion Beam applications have become routine in several areas. The Focused Ion Beam mill has become a standard for probe point creation and design rewiring (see Chapter 7). It has also become a preferred tool for cross-section preparation, both for SEM and TEM (see Chapter 9). In addition, focused ion beam technology provides very high sensitivity chemical analysis in SIMS (see chapter 14). Imaging is, therefore, not the most significant application of FIB[3]. The FIB spot sizes are significantly larger than for a SEM, making spatial resolution poorer. However, imaging is an important part of its applications because it allows navigation and precise location of the beam for milling. As with the interaction of an electron beam with a sample, the interaction of an ion beam with the sample surface causes a wide variety of events. The primary application of the focused ion beams is the sputtering which occurs. A variety of species, neutrals, positive ion and negative ions are created. The incident ions can also be backscattered or generate secondary electrons or be implanted in the sample. Just as with the SEM, a net charge imbalance usually results from the sum of these interactions. This charge imbalance causes much the same problems for imaging as in SEM. The most common signal used for imaging is the secondary electrons. Secondary electrons are typically detected just as in a SEM. Contrast mechanisms tend to be more material oriented than topography

Figure 10.11. FIB image of an aluminum pilot wafer after repeated scanning is shown.

oriented as in the SEM. The one significant advantage FIB secondary electron imaging has over SEM is the use of channeling effects. Channel effects[4] cause grain structures in material to be much more easily observed as shown in figure 10.11. Secondary ion imaging is also possible but is less common. Much of the discussion for SEM applies equally well to FIB since both are interactions of raster charge particle beams with the surface.

10.5 TRANSMISSION ELECTRON MICROSCOPY

TEM has been a useful tool for failure analysis activities for many years. It has been the very high-resolution option of choice in microscopy, particularly in viewing cross-section. TEM is performed by flooding a relatively large area with an electron beam. The image is formed and controlled by a series of lens and aperture, above and below the sample. The electron optics are capable of magnifications exceeding 500,000X. The image is formed directly on a phosphor screen, film or CCD camera.

As feature sizes shrink, resolution and charging artifacts limit SEM cross-section inspection, TEM utilization becomes more common. TEM requires preparation of extremely thin samples to allow electron transparency. As described in Chapter 9, the sample preparation for TEM can be tedious and difficult. However, FIB applications are significantly reducing the difficulty and time required for TEM sample preparation. Transparency is also improved through the use of much higher energy electrons than in a SEM, typically 100-300KeV. The higher energy beams are also less diffraction limited. In addition to cross section analysis, some growth in plan view TEM[5] has occurred as an option for viewing parallel polished devices. In plan view TEM, the device is thinned from top and bottom to produce and electron transparent sample containing the areas of interest.

TEM contrast is more straightforward than SEM contrast. Contrast is based only on the atoms in the area imaged and their ability to deflect the high-energy electron beam. Thus contrast is based primarily on nuclear density. The most obvious contrast arises from atomic number differences. However, contrast can also be generated from like materials of differing density such as silicon dioxides. Like SEM, TEM supports chemical analysis methods particularly EDS and EELS (Electron Energy Loss Spectroscopy). The spatial resolution of these analytical techniques is very high, particularly in STEM (Scanning Transmission Electron Microscope) where is small spot is generated.

10.6 SCANNING PROBE MICROSCOPY

SPM (Scanning Probe Microscopy) refers to a series of applications developed from AFM (Atomic Force Microscopy). Historically, AFM was derived from the Scanning Tunneling Microscope (STM), which was developed in the early 1980's at IBM Zurich[6]. Most of the SPM techniques involve moving a very sharp tip over the surface in a raster pattern and measuring displacements of the tip to maintain some parameter constant. The STM scans a surface while forcing a constant tunneling current to the surface and measuring displacement of the probe. STM does not

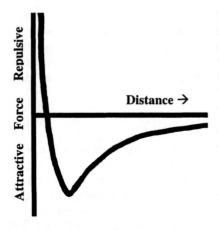

Figure 10.12. Van der Waals forces are very weak attractive forces at long distances. As the distances becomes shorter, the force increases but becomes repulsive at very short distances.

perform well on insulators, which is a major drawback in semiconductor applications. The AFM provides a topographical image of the device by rastering a tip while maintaining constant interaction force between the tip and the surface. The interaction with the tip is most often describe as by the van der Waals forces, which are the interaction forces between matter which only become significant at very short distances as shown in Figure 10.12. AFM is particularly important because most of the derivative techniques use AFM to build a topographic map of the sample and to maintain a particular spacing between the sample surface and the tip during measurements (see Figure 10.13). The AFM image can be generated prior to or coincident with the derivative application. A multitude of measurements, covering a full range of mechanical[7], physical[8], thermal[9], electrical[10-14], optical and mechanical[15-16], can be made with various tips as described below with very high spatial resolution. Innovative tip construction is the key factor for making the various measurements.

The central elements of all of these systems are a probe tip on a cantilever, raster generation, height detection and feedback. The probe tip is obviously the key part of all of these microscopes. Probe tips vary a great deal depending on the sample to be inspected and the form of SPM being applied. The probe tip is generally an integral part of a cantilever. Raster generation is typically done with piezoelectric crystals to generate extremely precise x-y movement. This movement is very accurate but the piezoelectrics raster pattern is usually limited to a few hundred microns on a side. Movement can be generated for either the sample or the tip. The systems must also be able to measure the height or deflection of the probe tip. This is most commonly done with a laser (see Figure 10.14). The laser beam is incident on the back of the cantilever and is deflected by

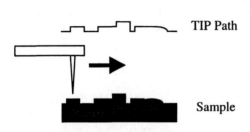

Figure 10.13. In SPM techniques, the tip is moved across the sample, typically at a fixed distance above the same.

changes in the height of the cantilever. The position of the reflection of the laser of laser beam is detected. The height information is used to provide feedback into the system. The deflection of position of the cantilever is commonly fed back to a height control mechanism, typically a piezoelectric. The feedback is used to maintain a constant height or deflection of cantilever. The z-movement is then used to map the contour of the sample.

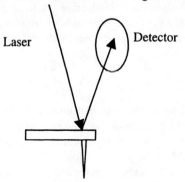

Figure 10.14. The typical SPM application uses a laser beam reflected from the top of the cantilever onto a detector.

AFM has several modes of operation. In contact mode, the probe tip actually is in the repulsive region of the van der Waals force regime and is touching the surface. The most common contact mode is the constant-force mode. The force of the tip on the sample is maintained at a constant point using feedback from the deflection of the cantilever. The topographical map is generated from the adjustment in sample height required to maintain a constant force.

AFM can also operate in a non-contact mode where the height of the tip above the sample is maintained in the attractive region of the van der Waals force curve. The probe tip is vibrated, typically on a stiff cantilever, near its resonant frequency (usually 100-400 kHz) with an amplitude of tens or hundreds of Angstroms. The resonant amplitude and frequency are both impacted by the interaction with the surface of the sample. Either the frequency or amplitude can be maintained constant through feedback from the cantilever deflection. This AC method is particularly important for other measurements since other AC measurements can be made simultaneously at other AC frequencies and sorted out use phase-lock amplification

Techniques for measurement of voltage, current, capacitance and other properties have been developed. All of these techniques continue to use the AFM capability to create physical images on which to overlay the electrical or other physical property, which is being measured. In addition, the AFM imaging is used to adjust the height of the detector during measurement. This can be a real time adjustment or it can be used as a two-part process where the contour is measured first and then the measurement is made. These techniques provide the same type of diversity of applications, which have made optical beam and scanning electron beam technology so important to the semiconductor industry. However, just as with optical microscopy and SEM, AFM imaging[15] is expected to remain an important part of the SPM set of tools. AFM provides a significant improvement in spatial resolution compared the SEM. It suffers from some major limitations. Slow rastering eliminates the possibility of real time imaging, which is critical for inspection of repetitive structures such as memory. The slow raster compared to a

SEM also limits throughput. In addition, the field of view is limited to the capability of the piezoelectric drivers, which can be extended[17]. In spite of these significant limitations, AFM is likely to become a significant inspection technology as resolution requirements become more demanding. The capabilities of simultaneous measurement of electrical, mechanical and optical properties will enhance its usefulness.

10.8 FUTURE CONSIDERATIONS

It can be readily assumed that the size of killing defects will continue to shrink with the feature sizes of IC's. This will continue to put pressure on the resolution of the primary defect documentation tools. Higher resolution tools can be expected to ultimately replace the optical microscope SEM as the primary tools. These tools are very familiar and popular and great efforts will made to extend their useful life. Optical microscopy is the only technique, which provides effective inspection of buried defects in the highly planarized devices. UV microscopes can be expected to use all available means to achieve optimum resolution: confocal, laser, immersion lenses etc. SEM can be expected to be the workhorse tool for many years. It has been used to develop many efficient and highly automated tools in a host of applications. Until replacement tools with higher resolution can become efficient and automated, they can be expected to remain research tools with focused applications. The present candidates for this task include the SPM and TEM. These tools have issues with sample preparation, throughput and the size of the area, which can be inspected.

In addition to resolution requirements, there is concern that a higher percentage of the defects will not be observable with microscopy (typically classified as NVD or no visual defect). For example, as the number of doping atoms used to form a junction becomes relatively small, statistical variations in the number of dopant atoms could impact the electrical performance of the transistor. This category of defect would be impossible to detect with standard microscopes. Tools such as the SCM (scanning capacitance microscope) could provide a powerful insight into localized electrical characteristics with very high spatial resolutions.

ACKNOWLEDGEMENTS

REFERENCES

1. Wagner L. An IR Microscopy Technique for Failure Analysis of Suspected Metallization Corrosion. Proceedings International Reliability Physics Symposium, 1980, 318.
2. Goldstein Joseph I, Newbury Dale E, Echlin Patrick, Joy David C., Fiori Charles, Lifshin, Eric. *Scanning Electron Microscopy and X-rayMicroanalysis*. New York: Plenum Press, 1981.
3. Wagner A, Longo P, Cohen, Blauner P. Focused Ion Beam Metrology. Journal of Vacuum Science and Technology B, 1995, 13(6), 2629.
4. Barr DL, Brown WL. Contrast Formation in Focused Ion Beam Images of Polycrystalline Aluminum. Journal of Vacuum Science and Technology B, 1995, 13 (6), 2580.
5. Okihara M, Tanaka H, Hirashita N, Nakamura T, Okada H, Hijikata Y, Shimoda K. Pin-Point Transmission Electron Microscopic Analysis Applied to Off- Leakage Failures of a Bipolar Transistor in 0.5

micron BICMOS Devices. Proceedings International Symposium for Testing and Failure Analysis, 1996, 207.

6   Binning G, Rohrer H. The Scanning Tunneling Microscope. Scientific American, 1985, 253, 50.

7   Yamashita H, Hata Y. Grains Observation Using FIB Anisotropic Etch Followed by AFM Imaging. Proceedings International Symposium for Testing and Failure Analysis, 1996, 89.

8   Balk LJ, Heiderhoff R, Koschinski P, Maywald M. Nanoscopic Evaluation of Semiconductor Properties by Scanning Probe Microscopes. Microelectronic Reliability, 1996, 36 (11/12), 1767.

9   Lai J. "Thermal Detection of Device Failure by Atomic Force Microscopy", IEEE Electron Device Letters, Vol. 16 (7), 312 (1995).

10   Campbell A, Cole EI, Dodd BA, Anderson RE. Magnetic Force Microscopy/Current Contrast Imaging: A New Technique for Internal Current Probing of IC's. Microelectronics Engineering, 1994, 24, 11.

11   Edwards H, McGlothlin R, San Martin R, U E, Gribelyuk M, Mahaffy R, Ken Shih C, List RS, Ukraintsev VA. Scanning capacitance spectroscopy: An analytical technique for pn-junction delineation in Si devices. Applied Physics Letters, 1998, 72 (6), pp. 698.

12   Hockwitz T, Henning A, Daghlian C, Bolam R, Coutu P, Cluck R, Slinkman J. DRAM Failure Analysis with the Force-Based Scanning Kelvin Probe. Proceedings International Reliability Physics Symposium, 1995, 217.

13   Nxumalo JN, Shimizu DT, Thomson DJ, Simard-Normadin M. High-Resolution Cross-Sectional Imaging of MOSFET's by Scanning Resistance Microscopy. IEEE Electron Device Letters, 1997, 18(2), 71.

14   Masters M, Slinkman J. Kaszuba P. Qualitative Kelvin Probing for Diffusion Profiling. Proceedings International Symposium for Testing and Failure Analysis, 1995, 9.

15   Cramer RM, Balk LJ, Chin R, BoylanR, Kammer SB, Reineke FJ, Utlaut M. The Use of Near Field Scanning Optical Microscopy for Failure Analysis of ULSI Circuits. Proceedings International Symposium for Testing and Failure Analysis, 1996, 19.

16   Duncan, WM. Near-Field Scanning Optical Microscope for Microelectronic Materials and Devices. Journal Vacuum Science and Technology B, 1996, 14, 1996.

17   Minne SC, Adams JD, Yaralioglu G, Manalis SR, Atalar A, Quate CF. Centimeter scale atomic force microscope imaging and lithography. Applied Physics Letters, 1998, 73 (12), 1742.

11

CHEMICAL ANALYSIS

Lawrence C. Wagner, Ph.D.
Texas Instruments Incorporated

Contamination is a major cause of failures in integrated circuits. Analysis of particles and thin film contamination is often essential to the corrective action process. Without identification of the chemical composition of the contaminant, it may be impossible to ascertain its actual source and eliminate or reduce contamination from this source. The total levels of contamination are frequently exceptionally small and concentrated in a very small volume. When selecting a method of analysis, several factors are frequently critical: the spatial resolution of the technique, the depth of material to be analyzed, the sensitivity of the technique employed, the type of chemical information provided by the technique and the ease of performing the analysis.

Semiconductor contamination typically occurs in the form of either particles or residual films. Particles may result in both shorts and opens on IC's. Shorts can occur due to bridging conductive particles or due to a particle blocking a metal etch process. Particles can also cause opens or high resistances when they fall on a contact and prevent proper filling of the contact with metal. Particles can also interfere with subsequent lithography steps in the wafer fab, resulting in a variety of defects. They frequently create a high spot in the device structure, which causes the area above it to be out of focus in subsequent lithography steps.

Thin films can reduce adhesion between materials or may prevent electrical contact. In the wafer fab, adhesion between metallization and dielectric is critical. Adhesion is also critical in many assembly processes including bonding and molding. Thin film contamination can prevent essential ohmic contacts in vias and contacts in the wafer fab. Thin film contaminants can also be barriers to important reactions such as solder wetting of leads.

Most analytical techniques can be viewed in terms of four critical elements: incident radiation; the physical interaction of the incident radiation and the sample; radiation flux from the sample which results from that interaction, either a new form of radiation excited by the incident radiation or an attenuation of the incident radiation; and a detector for the radiation flux from the sample. In fact, most of the physical analysis techniques can also be broken down into these respective parts.

11.1 INCIDENT RADIATION SOURCES

In selecting analytical techniques for semiconductor applications, the incident radiation sources are an important consideration. The incident radiation is usually a dominant factor in determining the spatial resolution of the technique. For chemical analysis, the spatial resolution can loosely be defined as the smallest area of the sample surface that can be analyzed with minimal interference from surrounding material. Spatial resolution is an exceptionally important requirement for analysis in semiconductors because of the small particles and very thin films to be analyzed. The very smallest particulates to be analyzed are often significantly smaller than the minimum feature size of the device. Most of the chemical analysis techniques and many of the physical analysis techniques, discussed earlier, rely on charged particle beams, i.e. electron beams and ion beams. This has occurred because charged particle beams can be tightly focused to a small spot, resulting in excellent spatial resolution. As in microscopy (Chapter 10), the trade off for the reduction in spot size is typically a reduction in the beam flux. This reduction in the radiation flux degrades signal to noise performance and sensitivity.

While resolution in the x- and y- dimensions is important, it is also important to consider depth resolution in the analysis of contaminants. This is a particularly important concept for the analysis of semiconductors because of their non-homogeneity at a submicron scale. The energy of the incident radiation will frequently determine how deeply the beam penetrates into the sample. Thus, the depth of resolution is directly impacted by the energy of the incident radiation. This dispersion of the incident radiation in the sample increases the interaction volume and degrades the spatial resolution. On the other hand, higher beam energies typically increase the interaction cross sections providing improved signal level. Higher beam energy can even be a critical enabling factor for certain types of interaction. For example, characteristic X-rays can not be generated with a higher energy than the incident e-beam. Another important element of analytical techniques for semiconductors is the common requirement that the techniques be as non-destructive as possible. Very high-energy beams can result in thermal and charging damage to a device. Thus, an appropriate energy of the incident radiation is a second critical concern.

Producing a small spot of radiation is largely based on the use of optics (light, e-beam or ion). Processing the source radiation through the optics results in a significant loss of intensity. In order to achieve a high signal to noise ratio, it is essential to have very bright sources. For light based technology, this typically is a laser. For e-beam techniques, this has led to the broad utilization of field emission (FE) sources. FE sources have essentially replaced thermionic

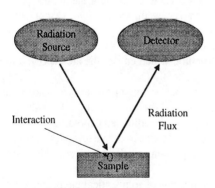

Figure 11.1. The key elements of chemical analysis techniques are the incident radiation source, the interaction with the sample, the resulting radiation flux and the detector.

emission systems in semiconductor applications due to higher brightness. Many ion beam techniques have developed accelerator-based techniques, e.g. HIBS (Heavy Ion Backscattering). Providing intense sources of radiation is a third critical concern in sources of radiation for chemical analysis in order to maximize a technique's sensitivity.

Thus, there are three important elements of the incident radiation. They are the spot size of the radiation, the energy of the radiation and the intensity of beam. They have a direct impact on the spatial resolution, depth of analysis and resulting signal level.

11.2 RADIATION-SAMPLE INTERACTION

When radiation interacts with a sample, it is very common for a large number of events to occur. For example, when an electron beam interacts with a sample, the interactions include the generation of secondary electrons, backscattered electrons, X-rays, and Auger electrons (see Figure 10.7). Each of these interactions can produce a signal, which can be analyzed and each occurs with a different probability or cross section. High cross section events tend to provide a higher signal level and thus better sensitivity.

The type of chemical information provided is typically dependent on the type of interaction between the incident radiation and the sample. There are basically two types of information obtained in failure analysis: atomic and molecular. Atomic (or elemental) information defines what elements are present in the volume sampled. Molecular information defines in some form the chemical bonding between the elements or at least the oxidation state of the elements present. Interactions, which occur at the atomic level, tend to provide only atomic information. This is particularly true of interactions that involve only inner shell electrons. Oxidation state information can be generated when the outer shell or valence electrons are involved in the interaction. Chemical bonding information is most commonly obtained when the interaction impacts more than one atom at a time such as FTIR (Fourier Transform InfraRed) interactions with chemical bonds or as the sputtering of clusters of atoms in SIMS (Secondary Ion Mass Spectrometry).

11.3 RADIATION FLUX

Once a resultant radiation is formed, only a portion of it will get to the detector to form the signal. This can be viewed in two parts, one escaping from the sample surface and the other travelling from the sample surface to the detector.

The depth of analysis is usually controlled by one of two factors, the depth of penetration of the incident beam as discussed above or the escape probability of the radiation flux from the sample as a function of sample depth. For example, in EDS, the depth of electron beam penetration controls the analysis depth. The radiation flux exits the sample with a minimum of inelastic interactions (interactions which cause a loss of part of the energy of the flux) and the entire interaction volume is analyzed. On the other hand, for Auger analysis depth, using the same incident radiation, analysis depth is controlled by the escape cross section of the Auger electron. Auger electrons formed in most of the interaction volume can not escape without inelastic interactions and a loss of their characteristic energy.

The radiation flux from the sample surface is typically directional in nature. In some cases, the direction of the radiation flux is determined by the direction of the incident radiation flux. A very simple example is reflection microscopy where the angle of incidence is equal to the angle of reflection. With the more complex chemical interaction, the radiation flux is typically most intense perpendicular to the surface. The variation with angle is frequently defined by a "cosine law", i.e. the flux is proportional to the cosine of the angle from the sample surface. This directional nature of radiation flux frequently determines what part of the resultant radiation can be effectively collected and measured. It also has an impact on the optimum location of detectors.

11.4 DETECTORS

In many cases, a range of detectors is available for the resultant radiation. The choice of detector can impact the sensitivity of the technique, the rejection of background signals and speed of analysis. The most dramatic difference in detectors is in characteristic X-ray analysis. Wavelength dispersive detectors provide much higher sensitivity and signal to noise but energy dispersive analysis provides much more rapid analysis. This speed of analysis has resulted in very high utilization of energy dispersive techniques with relatively little use of wavelength dispersive techniques.

11.5 COMMON ANALYSIS TECHNIQUES

Energy Dispersive X-ray Analysis (EDS) has become the primary chemical analysis tool for failure analysis and will be discussed in detail in Chapter 12. It provides a general-purpose analytical tool, which is uniquely suited to failure analysis requirements. Its primary advantage is that it is used in conjunction with a SEM, which is already available in the failure analysis laboratory. EDS has excellent spatial resolution with good sensitivity. Analysis can be performed quickly and easily. Interpretation of the spectra is relatively easy and can be automated. The quality of the spatial resolution arises from the small spot size of the SEM. The ability to move the electron beam enables the capability for dot mapping and line scanning. The depth of penetration for the incident electron beam is typically several microns. Since the escape cross section for X-rays is high, analysis is performed of the material up to that depth in the sample. This makes the surface a relatively small part of the sampled volume and results in X-ray analysis being of limited value for very thin surface films. The sensitivity of the technique is on the order of 0.1% for most elements but less for light elements. The primary disadvantages of the technique are moderate sensitivity, especially for light elements, and limited applicability to surface analysis. It provides atomic information and is not capable of providing molecular information.

Wavelength Dispersive Analysis detects the same interaction as Energy Dispersive Analysis. Hence, the type of data collected is identical, however, the different method of detection improves the resolution of X-ray peaks and yields improved signal to noise ratios. This results in improvement in sensitivity. The drawback is that the time for analysis is long relative to energy dispersive analysis.

EDS is the primary technique for chemical analysis because of its ease of use, low cost and availability as a SEM accessory. It is a powerful general-purpose tool but has deficiencies for several types of analysis. EDS analysis becomes limited for the study of surface interactions because of its relatively high depth of penetration. Auger often fills this gap. EDS sensitivity is moderate. Applications that require very high sensitivity are commonly performed on SIMS. Atomic information provided by EDS is usually adequate for analysis of inorganic materials. However, when molecular information is required to identify organic contamination, Microspot FTIR (Fourier Transform InfraRed) is commonly used. Auger, SIMS will be discussed in detail in Chapters 13 and 14. FTIR is discussed in more detail below.

Key Properties	EDS	Auger	SIMS	FTIR
Depth Analyzed	1-5 Microns	Surface Top 20 A	Surface	1-10mm
Sensitivity	0.1%	0.1%	PPM-PPT	PPM
Primary Information	Elemental	Elemental	Elemental/ Molecular	Molecular
Spatial Resolution	100A	150A	0.3-0.5 micron	2mm
Radiation Source	Electron Beam	Electron Beam	Ion Beam	IR Light

Table 11.1 The primary chemical analysis methods for failure analysis are summarized. EDS is the primary method due to its ease of use and availability on SEM. The particular strength that makes each of the others important is italicized.

11.6 MICROSPOT FTIR

EDS, Auger and SIMS have one common shortcoming for the analysis of organic contamination. They provide predominantly atomic composition information. However, TOF-SIMS (Time of Flight SIMS) is finding growing application for determining the composition of organic surface contamination (see Chapter 14). For most inorganic materials, atomic composition provides adequate information in order to identify contaminants very accurately. Organic contaminants, on the other hand, are widespread and cover a vast range of varied materials. They are extremely difficult to differentiate based on atomic composition information. In general, they are made up predominately of carbon, hydrogen, oxygen and nitrogen. Information from these techniques typically identifies the presence of carbon, oxygen and nitrogen. Hydrogen is not detectable by Energy Dispersive Analysis or Auger. The presence of other elements such as halogens or sulfur can provide important clues as to the precise organic compound that is present. However, this is often inconclusive since these elements are typically present at a low atomic percent if at all.

Infrared Spectroscopy provides a method of identifying the organic building blocks that are present in a contaminant molecule. This occurs through bending, stretching and vibration transitions within the chemical bonds. The organic molecules have well defined energy levels associated with the bending, stretching and vibration of each chemical bond. Transitions occur when infrared light, having exactly the energy required in order to cause a transition between these well-defined states, is absorbed by organic molecules. Each type of bond will have an absorption spectrum that is typical of that type of bond. The exact energy value will, however, be affected by what is attached to the atoms forming the bond. For example, a carbon oxygen double bond exhibits absorption in a narrow range. The exact energy of absorptions will depend on the group attached to the carbon. Thus, aldehydes where one hydrogen atom and another organic group is attached to the carbon will have slightly different absorptions from a ketone, which has two organic groups attached to the carbon atom. Similarly, the absorptions for an acid where an organic group and –OH group are attached to the carbon atom will be different. An initial interpretation of an IR spectrum is the identification of the types of bonds present. Typical identifications might include: carbon-carbon bonds (single, double and triple), aromatic groups (benzene derivatives), carbon-oxygen (single and double bonds), carbon-nitrogen bonds (single, double and triple), etc. The initial identification can also include more specific information such as the differences between aldehydes, keytones and acids.

Each bond in the organic molecule potentially provides a distinct set of absorption peaks. Since there are a large number of bonds in a typical organic molecule, the spectra of compounds of interest tend to be rather complex and difficult to interpret. Thus, comparison to known spectra is very important. Vast numbers of organic IR spectra are cataloged in libraries and can be automatically compared to the contamination spectra. This process is much like fingerprinting. If a spectral match is not found for a contaminant in a library, likely sources can also be analyzed. Thus, organic materials, with which the device is likely to have come into contact, are also analyzed and compared to contaminant spectrum. At times is may be necessary to chemically separate the components of a possible contamination source. For example, a polymer may have a number of components such as cross-linked resins, dyes, release agents and plasticizers. Components, which outgas, can be collected by heating the sample followed by IR analysis of components that come off at various temperatures. They can be further separated if required by gas chromatography. The component spectra can then be compared to the contaminant spectrum.

Infrared spectra are collect by passing infrared light through a sample and measuring the absorption of the IR light as a function of wavelength. Liquids or emulsified solids have been analyzed by holding the material between crystals, which do not absorb in the IR. Similarly, gases contained in a vessel with IR transparent walls can be analyzed by simply measuring absorption of IR light passing through the vessel. Historically, IR spectra were obtained by monochromating the IR light signal and comparing the intensity as function of wavelength to that of the IR source through a comparable optical path (without the sample). In FTIR, the light is analyzed with an interferometer. The resulting signal is the Fourier transform of the IR spectrum. The result is converted back to the spectrum using a Fast Fourier transforms. Since all of the wavelengths of the spectrum are analyzed at one time, the FTIR method is significantly faster.

The techniques described above provide IR spectra of macroscopic samples. In microspot FTIR[1-4], a microscope is used to create a very small (as small as several microns) analysis area. This allows analysis of small particles and very localized contamination. By integrating spectra over long times, spots of organic contamination, which are smaller than the microscope field of view, can be analyzed at times. However, spatial resolution remains a concern for some applications. The relatively limited spatial resolution, compared to wafer fab particles, has led to Microspot FTIR application being focused in assembly areas. Since potential organic contaminants are wide spread in the assembly areas, this is an excellent match for a range of assembly related issues.

11.6.1 Micro-Raman Spectroscopy

Chemical bonding information similar to that found with IR Spectroscopy can be obtained in Raman spectroscopy. In Raman scattering, a monochromatic light beam is used, usually in the visible region. The light undergoes inelastic scattering in the sample. The energy associated with a vibration in the material can be added or subtracted from the energy of the monochromatic light source. Monochromatic light can be more easily focused into a small spot and scanned. Due to these advantages, Micro-Raman has several unique applications. The phonon interactions show dependence on crystallinity and mechanical stress. Taking advantage of these dependencies, it has been successfully used, for example to measure crystallinity[5] in silicon and localized stresses[6] in dielectrics and silicides. Micro-Raman can also be combined with near-field microscopy[7] to provide very high spatial resolution chemical analysis. On the negative side, the Raman scattering signal is weak and must be separated from the intense signal at the wavelength of the incident light, making spectra acquisition difficult.

11.7 OTHER TECHNIQUES

A number of other analytical techniques are occasionally required in failure analysis. Some of these fill specific niches in analytical capability.

XPS[8-10] (X-ray Photoelectron Spectroscopy, also known as ESCA or Electron Spectroscopy for Chemical Analysis) observes electrons that are ejected from atoms due to an interaction with monochromatic X-rays. Elemental characterization is by binding energy (the X-ray energy minus the kinetic energy of the electron). The binding energy is characteristic of elements. Molecular information is obtained from energy shifts with oxidation states. The shifts are much more pronounced than with Auger. In addition, the X-ray interaction is primarily with outer shell electrons whose energies are most impacted by the chemical surroundings. This is also a surface technique, since as with

Auger, the probability of electron escape without energy loss from below the surface is low. In fact, Auger peaks are also observed in the analysis of electron energies as background. The primary limitation of the technique is focusing of the incident X-ray beam to a small spot. Significant progress has been made in columnating X-ray beams. However the spatial resolution remains inadequate for many failure analysis applications. The technique may be used in conjunction with Auger analysis in order to compliment the elemental information of Auger with oxidation state information of XPS.

X-ray Fluorescence (XRF), like EDS, uses characteristic X-rays. In XRF, X-rays are used to generate the initial inner shell electron vacancy as opposed to an electron beam in EDS. In TRXRF (Total Reflection X-ray Fluorescence), the X-ray beam contacts the sample at a very shallow angle. This results in a shallow and well-controlled depth of analysis. The technique provides higher sensitivity than EDS and is ideally suited for the analysis of metallic contamination in relatively large areas. The spatial resolution is not adequate for most failure analysis applications.

Rutherford Backscattering[11] or RBS looks at backscattered high-energy ions. Typically the ion beam is MeV He or H. The technique provides excellent quantitative analysis to measure the stoichiometry (the chemical composition or ratio of atoms of different types in a material). In a channeling mode, RBS provides a useful tool for determining how well ordered a crystalline structure is. HIBS (Heavy Ion Backscattering) is an accelerator version of the technique using ions with higher atomic numbers.

While spatial resolution is often a primary concern in failure analysis, there are occasions when this is not the case. When non-localized ionic contamination is suspected, Ion Chromatography provides an excellent analysis options. Ions are typically extracted from the sample, either a group of packages or wafer. The most common method is boiling water extraction. The extract is typically concentrated and then analyzed on an Ion Chromatography column. Both anions and cations can be analyzed on different types of columns.

Several techniques[12] have found niche applications for assembly related issues. Residual gas analysis (RGA) provides a method for assessing the composition of the gas in a cavity package. The package cavity is ruptured in a vacuum. The package may require some mechanical thinning to facilitate the rupture. The gas escaping from the package is analyzed, typically with a quadrupole mass spectrometer. RGA is most commonly used to detect moisture in the package. The presence of air, rather than the gas sealed in the cavity (typically nitrogen), can be used to confirm a leak. Atomic Absorption Spectrometry (AAS) is commonly used to assess the purity of metal and is particularly used in bonding wire analysis. AAS provides excellent sensitivity for low concentrations of metallic impurities in bonding wire. In addition to chemical analysis techniques, mechanical analysis can play a significant role in the analysis of polymer related issues. These techniques provide an excellent method to assess the cured state of the polymers. For failure analysis, this applies predominantly to mold compounds used in plastic packages although die-attach materials and other polymeric materials can also be analyzed. Thermal Gravimetric (weight change during heating) and Thermal

Mechanical Testing (coefficient of thermal expansion) are both used to characterize polymers.

11.8 CONCLUSION

Obviously, one challenge of failure analysis is to select the analytical techniques that are best suited to a particular situation. EDS is typically the first choice due its ease-of-use and availability. Auger, SIMS and FTIR provide a solution in most instances when EDS can not provide an adequate analysis. These core techniques resolve most of the analytical challenges in failure analysis.

The future challenges of chemical analysis are not dissimilar to those of failure analysis in general. Improvements is spatial resolution is critical. This need extends to characterization of smaller IC features as well as to smaller particles. The spatial resolution issue is not restricted to the x-y planes. Better depth profiling is also critical, as junctions become shallower. A second area of concern is sensitivity. This is also driven by the reductions in feature sizes. As features become smaller, lower levels of contamination will be able to disrupt device operation. In addition, there will inevitably be new niche requirements.

REFERENCES

1 Shala FJ. Applications of FTIR Microspectroscopy in Contamination Analysis. Proceedings International Symposium for Testing and Failure Analysis, 1992, 175.
2 Kudva S, Knudson E. Analysis of On-line Organic Microcontaminants in Semiconductor Assembly Plants. Proceedings International Symposium for Testing and Failure Analysis, 1986, 1.
3 Peters DC. Microscopic FTIR for Failure Analysis. Proceedings International Symposium for Testing and Failure Analysis, 1985, 177.
4 Martin FC, Reyes WA. Fourier Transform Infrared Spectroscopy: Its Application to Failure Analysis. Proceedings International Symposium for Testing and Failure Analysis, 1984, 53.
5 Perkowitz S, Seiler DG, Duncan WM. Optical Characterization in Microelectronic Manufacturing. Journal of Research of the National Institute of Standards and Technology, 1994, 99, 605.
6 DeWolf I, Pozzat G, Pinardi K, Howard DJ, Ignat M, Jain SC, Maes HE. Experimental Validation of Mechanical Stress Models by Micro-Raman Spectroscopy. Microelectronic Reliability, 1996, 36 (11/12), 1751.
7 Duncan, WM. Near-Field Scanning Optical Microscope for Microelectronic Materials and Devices. Journal Vacuum Science and Technology B, 1996, 14, 1996.
8 Peignon MC, Clenet F, Turban G. Contact Etching Process Characterization by Using Angular X-Ray Photoelectron Sprectroscopy Technique. J. Electrochemistry Society, 1996, 143 (4), 1347.
9 Iwata S, Ishizaka A. Electron Spectroscopic Analysis of the SiO_2/Si System and Correlation with metal-oxide-semiconductor Device Characteristics. Journal Applied Physics, 1996, 79(9). 6653.
10 Kelley MA. Using ESCA in Failure Analysis: Some Recent Developments. Proceedings International Symposium for Testing and Failure Analysis, 1984, 35.
11 Diebold A. Materials and Failure Analysis Methods and Systems Used in the Development and Manufacture of Silicon IC's. Journal American Vacuum Science Technology, 1995, 2768.
12 Wagner L. "IC Package Reliability Testing". In Characterization of Integrated Circuit Packaging Materials, Thomas M. Moore and Robert G. McKenna eds. Stoneham, MA, Greenwich, CT: ButterWorth Heinemann Manning, 1993.

12

Energy Dispersive Spectroscopy

Phuc D. Ngo
ST Microelectronics

Energy Dispersive Spectroscopy (EDS, also sometimes called EDX or Energy Dispersive X-ray Analysis) is the most frequently used chemical analysis tool in failure analysis. It has some very significant advantages. It is used as an attachment to the SEM (Scanning Electron Microscope), which is readily available in every failure analysis laboratory. Analysis is performed in minutes. The spectra are easily interpreted. Spatial resolution is good. It also has some limitations as an analysis tool. Sensitivity is limited to concentrations on the order of 0.1% in the sampled volume. A second limitation is the sampled volume is relative large compared to the thickness of semiconductor thin films and deep submicron particles. A final limitation is that it provides strictly atomic information as opposed to molecular. These limitations are addressed by the three other key chemical analysis techniques, which will be discussed in the next three chapters.

While much of the EDS work is performed in SEM's, EDS systems also function on TEM (Transmission Electron Microscope) with some significant advantages in terms of interaction volume. They are also used with the SEM in dual column FIB (Focused Ion Beam) systems.

12.1 CHARACTERISTIC X-RAY PROCESS AND DETECTION

The interaction of an electron beam with a sample generates a host of useful radiation- types for analysis including secondary and backscattered electrons, characteristic X-rays[1] and Auger electrons. The characteristic X-ray is used for EDS. The characteristic X-ray generation process (Figure 12.1) is initiated with the ejection of an inner shell electron to form a vacancy. From this excited state, an upper shell electron drops into the inner shell vacancy. An X-ray is generated with an energy equal to the difference between the energies of the electron shell. X-rays are generated in most of the interaction volume of the electron beam with the sample. Other radiation sources can be used to generate the same initial excited state; for example X-rays are used in TXRF to generate the same characteristic X-rays. An electron beam is preferred in failure analysis because of the smaller beam spot sized and resulting better spatial resolution.

The nomenclature for various X-rays generated is as follows. The initial nomenclature is taken from the electronic shell of the initial vacancy created by the electron, using K, L, M, and N as shell names. A Greek letter is used the designate the number of shell above the vacancy from which the replacing electron falls, α, β, γ, δ (Figure 12.2). Fine structure in the electronic shell results in structure in the X-rays. This fine structure is denoted by subscript numbers, e.g. $K_{\alpha 1}$.

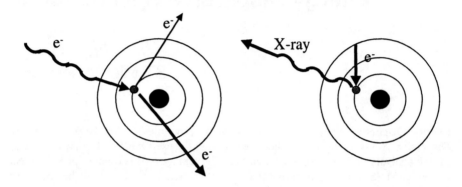

Figure 12.1. X-ray generation occurs in two phases, inner shell vacancy generation (left), followed by the filling of the vacancy from a higher shell with X-ray generation (right).

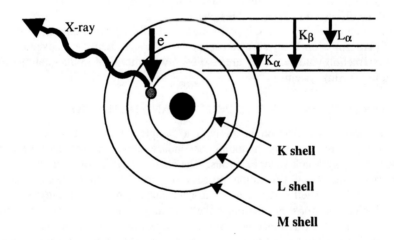

Figure 12.2. X-ray nomenclature based the shell of the hole and the number shell above it from which an electron comes to fill the vacancy is demonstrated. A Kβ transition is shown in the diagram.

The X-rays are detected using an energy dispersive analyzer. This is typical a SiLi detector (Lithium Drifted Silicon). Germanium detectors can also be used. The X-rays are used to create an electron pulse where the height of the pulse or number of carriers generated is proportional to the energy of the X-ray. Hence, the detection method is commonly referred to a pulse height analysis. The X-ray spectrum can also be measured by Wavelength Dispersive Spectroscopy (WDS). In this case, the X-ray spectrum is analyzed by diffraction through crystals. Since X-rays are very precisely diffracted by the crystal according to Bragg's Law, resolution of wavelengths is much better. The drawback of this technique is that the X-ray intensity is measured essential one wavelength at a time. This leads to a long time to acquire spectra. In addition, generation of a complete spectrum often requires several crystals to be used, further adding to the analysis time.

12.1.1 Electron Beam Adjustments

The three key characteristics[2] of the electron beam are the spot size, accelerating potential and beam current. Spot size is not as critical a factor for EDS as for SEM. This is because the sampled volume for the EDS is most of the interaction volume of the electron beam. This is a large volume relative to the size of the electron spot on the surface. Thus decreasing the spot size does not have a comparable effect on the analysis area. Variation of the accelerating potential of the electron beam has several impacts on X-ray generation. Higher energy electrons provide more efficient X-ray generation and can excite higher energy X-rays. X-rays are not created with an energy higher than the incident electron energy. On the other hand, increasing the accelerating potential causes the penetration of the electrons into the sample to increase, enlarging the sampled volume. This degrades spatial resolution and usually increases the background signal. Beam currents typically used for high resolution SEM do not provide enough X-rays for efficient spectra acquisition. Beam current must be increased to generate enough X-rays for efficient acquisition. Factors which limit the amount of beam current, which can effectively utilized include detector dead time, sample heating and charging. The dead time of the detector is the percentage of time during which the detector is not measuring X-rays. A number of factors contribute to dead time including signal processing time in the preamplifier and rejection of multiple X-ray events. Since each X-ray is detected as separate event or pulse height, the systems must have software to reject multiple X-ray events. Typically dead times on the order of 60% are desired.

12.1.2 Detector Geometry Considerations

While several of geometric factors are fixed by the EDS detector location in various SEM's, it is worthwhile understanding these issues. The solid angle of collection is important because the X-rays are generated at all angles from the sample surface and are not actively collected. In order to collect the X-rays efficiently, a large solid angle is required. This can be achieved by increasing the area of the detector and moving it closer to the sample. The various trade-offs made in the area of the detector and its distance from the sample determine the actual location of the detector in SEM's.

The incident angle of the electron beam, usually the sample tilt angle, also impacts EDS by altering the interaction volume (see Figure 12.3). As the tilt angle is increased, the interaction volume is brought somewhat closer to the surface. This is also the basis of TRXRF where the incident X-ray beam is generated at a glancing angle to provide analysis with a shallow interaction volume.

The take-off angle is the angle between the sample surface and the line from the spot under analysis and the center of the detector. The maximum X-ray intensity occurs normal to the surface of the sample or a 90-degree take-off angle. Since the X-ray intensity varies as the cosine of the take-off angle, the X-ray signal drops off rapidly at small take-off angles. Hence low take-off angles must be avoided. It should be noted that the take-off angle is dependent on the tilt angle. Detector location is thus usually optimized for typical combinations of tilt and take-off angles. Typical tilt and take-off angles are in the 45-60 degree range. In general, low tilt angles should be avoided.

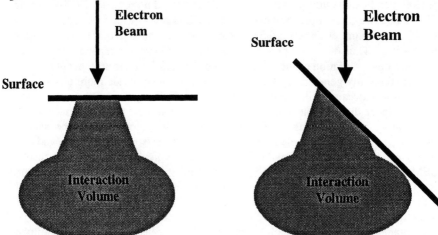

Figure 12.3. The effect of sample tilt on the distance of the interaction volume from the surface is demonstrated.

12.1.3 Detectors and Detector Windows

X-rays are detected by electron-hole generation in the SiLi or Ge detector[3]. Bias across the detector results in a current flow, which is proportional to the number electron-hole pairs generated. The number of electron-hole pairs generated is proportional to energy of the incident X-ray. This occurs with approximately 3.8-3-9 eV of X-ray energy dissipated per electron-hole pair generated. Since there are competing events in the detector such as heat generation, there is a statistical variation in the number of electron-hole pairs generated for identical X-rays which results in a normal distribution of peak heights. The X-ray events are sorted by peak height or pulse height as measured by the carrier collected during each event.

Other factors in the detector also impact peak shape. A dead zone typically occurs in the top surface of the detector. This dead zone occurs because the lithium introduced is not able to totally neutralize surface defects in the silicon. This results in traps near the surface. The traps combine with the carriers to reduce the current between the electrodes. This results in a low energy tail superimposed on the normal distribution of the peak.

Another feature of the SiLi detector is escape peaks. If a silicon Kα X-ray is generated by the incident X-ray and escapes from the detector, the total energy available for electron-hole generation will be smaller by the 1.74 KeV of the Silicon X-ray. Escapes normally occur at the surface of the detector. Escape peaks are normally on the order of few percent of the real peak and can be corrected for by subtraction from the spectra.

The detector environment is also critical. The detector must be maintained at liquid nitrogen temperatures to reduce shot noise and it must also be maintained in a clean environment. Typically the vacuum for the detector is maintained separately from the sample chamber and requires a window to protect the detector. The standard window is a beryllium window which transmits higher energy X-rays but absorbs lower energy X-rays. This limits detection to elements with atomic number 11 (Na) and above. Thin windows are also available which are transparent to the lower energy X-rays. It is also possible to remove the window during analysis for light element analysis windowless detection.

12.2 QUANTITATIVE ANALYSIS

EDS is commonly used for qualitative or semi-quantitative analysis. Modern systems allow computer manipulation of the spectra to generate quantitative results. A first step is to integrate the area of the EDS peaks, essential the number of X-rays in the peak. Background signals such as Bremsstrahlung[1] must first be subtracted out. Bremsstrahlung radiation is X-rays, which are created by the inelastic interaction of the electron beam with nuclei. This radiation forms a continuum because the interactions are not defined by quantum levels. The energy of Bremsstrahlung X-ray is higher for very close interactions with the nucleus. Since close interactions are less likely than distant interactions, the continuum is more intense at low energies than high energies.

In addition to background subtraction, several corrections known as ZAF corrections must be made. "Z" in ZAF refers to the atomic number effects. This includes excitation efficiency, fluorescent yield and detector efficiency. The "A" correction is for absorption of the X-rays exiting the sample. The "F" correction is for fluorescence that is caused by X-rays generated from other elements present as well as from the continuum background.. Use of the ZAF correction is highly dependent on the geometric factors discussed above: incident electron angle and take-off angle. This makes the corrections for irregular objects such as particles less meaningful. In addition, it is generally assumed that the X-rays come from a point in the sample and the sample is assumed to be homogeneous. This makes analysis where different layers are sampled less accurate. ZAF factors have historically been applied to k-ratios, ratio of the net peak counts to the peak counts for the standard. Standard-less analysis can be performed by estimating the intensity of know standard from the set-up conditions. For routine analysis, calibration curves can be established for the materials studied. A good example in the semiconductor industry is the level of phosphorus in silicon dioxide. If intensity ratios are obtained for several standards of know composition, the resulting calibration can generate relatively accurate results (Depth distribution of the phosphorus will impact the results. Thus it must be assumed that the distributions in the sample and standards are similar).

The most common failure analysis applications of quantitative analysis by EDS are detection of variations in composition from a good device to a failed device. This could be a higher or lower level of phosphorus in oxide or variations in lead to tin ratios in solder. In making such comparisons, the use of the same conditions for standards if used and the samples of interest is essential.

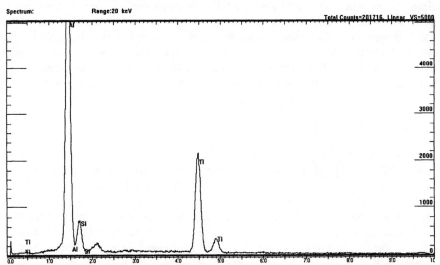

Figure 12.4. Spectral analysis indicates the elements present in the defect shown in Figure 12.5.

12.3 SAMPLE CONSIDERATIONS

As discussed above, geometric considerations are very important. Tilt angle and take-off angle are very critical and interrelated. Since beam current is high for EDS and accelerating potential is typically medium to high, charging effects are quite significant. While sample coating is, in general, not desirable, it may be required in some cases for EDS. Typical coatings are sputtered carbon or metals which are unlikely to generate EDS peaks in an area of interest.

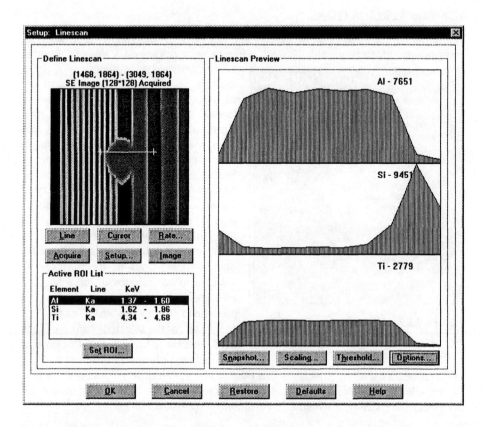

Figure 12.5. Linescans of a lithography defect is shown. The graphs to the right indicate the elemental concentrations along the line drawn on the defect.

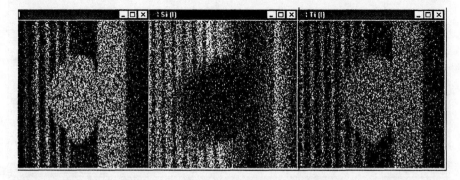

Figure 12.6. Dotmaps of the specific elements in the defect area indicate the concentration of the selected elements (Al, Si, Ti from left to right).

12.4 EDS APPLICATIONS

While the composition of some defects appears obvious and simple, EDS analysis can provide information which offers different perspectives on a defect. Figures 12.4-6 shows typical outputs of EDS analysis of a defect. The peaks in the EDS spectrum (Figure 12.4) are usually labeled by the EDS software. Unfortunately, the software may also label a secondary peak as an element, which cannot be part of the process. Peaks from known elements can be masked out with operator intervention. An experienced operator would also be able to re-label secondary peaks which most software may label incorrectly.

Another output format is the linescan. Specific peaks from the spectrum can be selected to be included in the scan. The electron beam is scanned along a predefined line in the SEM image. The count rate for the selected channels of the EDS detector (corresponding to elements selected) are plotted as a function of the electron beam location. Linescans are used to determine qualitative concentrations of the selected elements across a defect (Figure 12.5). It can also be useful for predicting the location at which a defect should be cross-sectioned.

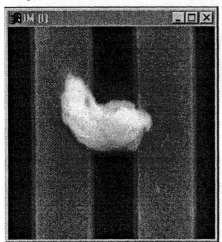

Figure 12.7. SEM image showing a stainless steel particle causing a short between two nodes.

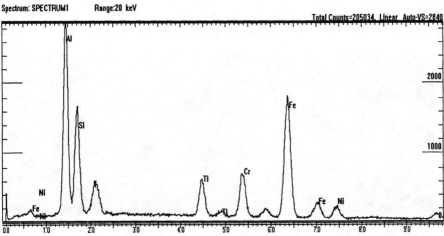

Figure 12.8. Spectral analysis indicates the elements present in the defect shown in Figure 12.7.

Dotmaps provide a two dimensional alternative to linescans. A dot is displayed at each pixel for which the EDS count rate exceeds a specified value for the selected element. Thus the primary area(s) where an element is present are highlighted. While more qualitative than line scans, dotmaps provide an excellent visualization of contamination location. Dotmaps of the defect area for the key elements detected are shown in figure 12.6.

Figure 12.7 shows a typical defect found in failure analysis. It is critical to identify and localize the elements present in the particle in order to determine its origin. Understanding all the elements detected is key to linking the defect to the contamination source. As shown in Figure 12.8, various foreign elements are detected in the defect. Chromium, iron, and nickel are not present in wafer fab processed materials. However, these elements are major components of stainless steel, which is present in numerous wafer fab tools, which are potential contamination sources in the clean room environment.

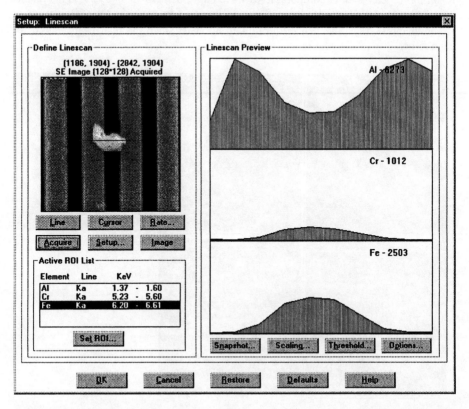

Figure 12.9. Linescan of a particle defect. The graphs to the right indicate the elemental concentration along the line drawn on the defect.

Linescans (Figure 12.9) and dotmaps (Figure 12.10) yield additional information about the defect. It is indicative of where the defect has the most volume and the distribution of the concentration. Gaps in the dot pattern can indicate the possible location of other contaminants to guide further analysis.

12.5 FUTURE CONSIDERATIONS

As feature sizes shrink, EDS becomes more limited due to a combination of factors: large sample volume, poor signal to noise and modest sensitivity. Much effort has been expended in attacking the issue of sampled volume by decreasing the accelerating potential. This reduces the sampled volume but limits X-ray generation to those energies lower than the incident electron.

A relatively direct potential replacement for EDS is microcalorimetry[4,5]. The energy of the X-ray is detected as heat dissipated in a thermal detector held at liquid helium temperatures. Noise and competing events are significantly reduced resulting in peak widths comparable to wavelength dispersive analysis. This results in greatly improved sensitivity without an increase in analysis time. Rejection of simultaneous X-rays and dead time remain an issue since the heat is measured for each X-ray.

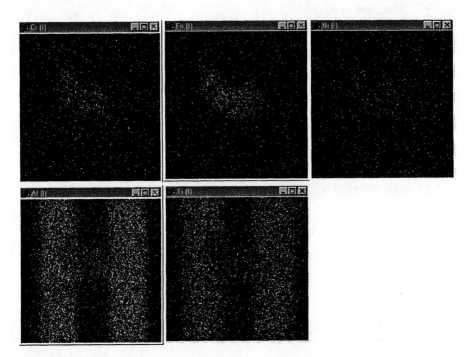

Figure 12.10. Dotmaps of the elements (Cr, Fe, Ni across top row) in the particle area easily show the foreign elements that are constituents of the stainless steel particle. Al and Ti (bottom row) are normal elements in the metallization stripes.

REFERENCES

1   Krane, Kenneth. *Modern Physics*, pages 74 and 214.. John Wiley and Sons, 1983.
2   Goldstein Joseph I, Newbury Dale E, Echlin Patrick, Joy David C., Fiori Charles, Lifshin, Eric. Scanning Electron Microscopy and X-ray Microanalysis. New York: Plenum Press, 1981.
3   Kittel, Charles. *Introduction to Solid State Physics*, page 303. John Wiley and Sons, 1986.
4   Silver E, LeGros M, Madden N, Beeman J, Haller E. High-Resolution, Broad-Band Microcalorimeters for X-ray Microanalysis. X-ray Spectrometry, 1996, 25, 115.
5   Wollman DA, Irwin KD, Hilton GC, Dulcie LL, Newbury DE, Martinis JM. High-Resolution, Energy-Dispersive Microcalorimeter Spectrometer for X-ray Microanalysis. Journal of Microscopy, 1997, 188, 196.

13

AUGER ELECTRON SPECTROSCOPY

Robert K. Lowry
Harris Semiconductor

Auger Electron Spectroscopy[1-6] (AES) is an essential tool for analyzing failed IC's where surface characterization is required. It provides elemental identification and, in some cases, chemical information about substances in highly localized areas in very near-surface regions of materials, thin films, or film interfaces. AES provides excellent spatial resolution, as low as 200 Angstroms diameter. It provides surface analysis of the uppermost 20-50 Angstroms of the sample surface. It has moderate sensitivity with detection limits in the 0.1-1.0% range. Depth profiling can be achieved by specimen sputtering.

The first part of this chapter reviews atomic electronic structure and describes the Auger electron process. Important considerations for sample analysis by AES as well as the elemental and chemical information provided are discussed. Applications of AES in failure analysis of IC's are presented

13.1 THE AUGER ELECTRON PROCESS

Pierre Auger (pronounced oh-zhay) first described the electron energy transition upon which AES is based in 1923. In AES, a beam of primary electrons bombards a solid specimen surface. The electron bombardment induces a number of energetic processes within the sample, including removal of electrons from the inner electronic orbitals of the atoms comprising the material. The result is an energy imbalance for each atom where an electron vacancy is induced. The energy imbalance is satisfied by electron rearrangement, where an electron from the next higher energy level decays into the electronic orbital vacancy. The result is emission of either an X-ray or an energetic electron. An electron emitted from an atom under these conditions is known as an Auger electron.

To understand the nature of Auger electron rearrangements, recall that atoms have nuclei surrounded by electrons occupying orbitals of discrete energy values. As an example, the electronic structure for the silicon atom is shown in Figure 13.1A. Moving outward from the nucleus the electron orbitals are designated by the principal quantum numbers 1, 2, 3, with letter designations K, L, M etc. Each of these principal energy levels or shells contains energy subshells. The K shell has one energy level (K_1) and can be occupied by 2 electrons. The L shell has 2 subshells, one (L_2) at one energy level and three (L_3) at a higher energy level. The L shell can be occupied by a total of 8 electrons, 2 in L_2 and 6 in L_3. The M shell has three subshells at three different energy levels and can be occupied by 18 electrons, i.e. 2, 6, and 10 electrons respectively in the three subshells. The electronic structure of shells with higher principal quantum numbers continues to grow in complexity. In silicon, the M level is only partially occupied with four electrons of equivalent energy known as the valence band

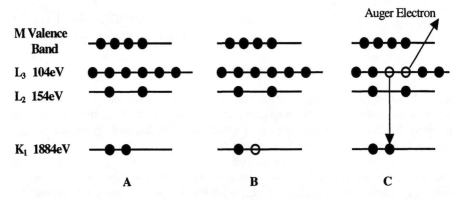

Figure 13.1. Electronic structure of silicon (A) is shown with the configuration after vacancy creation (B) and Auger Transition (C).

Figure 13.1 depicts an Auger process for a silicon atom. The process is initiated by bombardment with a beam of 2-25KeV primary electrons. Bombardment causes a number of secondary electron emissions. When a core shell electron is removed, an electronic vacancy is created as shown in Figure 13.1B for a K shell electron with binding energy 1844 eV. The resulting energy imbalance of the atom can result in one of two events, either emission of an X-ray (basis of Energy Dispersive Spectroscopy) or a radiationless transition of an electron from the L shell to fill the energy vacancy in the K shell. Where no radiation is emitted, the energy involved in this electron's transition is transferred to a second L shell electron, which is ejected from the L shell as shown in Figure 13.1C. Electrons ejected from atoms by this energy transfer process are termed Auger electrons. The energy of the Auger electron is defined by the energy difference of the electron states involved in the vacancy filling minus the energy loss in removing the Auger electron from the atom and the sample surface. In the example shown in Figure 13.1, the energy difference in the vacancy filling is 1844 eV − 104 eV = 1740eV and the energy loss in removing the Auger electron from the atom and then from the silicon surface is 104 eV + 17eV (work function of silicon) = 121eV. The ejected electron's energy of 1619 eV (1740 − 121) identifies it as coming from a silicon atom. This is the basis for elemental analysis and identification by AES. The transition shown in Figure 13.1 is defined as a KLL Auger transition based on the shell letter designators for the vacancy and two electrons involved in the transition.

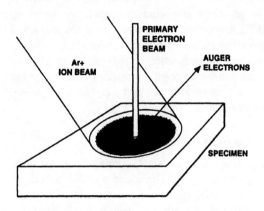

Figure 13.2. Sputtering is performed over a large area compared to the analysis spot.

Auger transitions can also involve vacancies in the level just below the valence level and valence shell electrons (designated by V rather than the shell letter). For example, in the silicon atom these are termed LVV transitions. Note that this type of transition can occur with the vacancies, which result from the KLL Auger transition. This results in cascading. Valence band electron transitions can provide information about the chemical state of an atom induced by its surroundings because the type and arrangement of nearest neighbor atoms influence the inherent energies of valence band electrons. This is reflected in shifts in position and shape of the Auger peaks.

13.2 AES INSTRUMENTATION AND CHARACTERISTICS

The primary components of an AES system are an electron gun for primary electron beam excitation and an electron energy analyzer. These are located in an ultrahigh vacuum sample chamber (1E-9 pascal) with a stage for sample mounting and movement. The electron gun generally operates in the 2-25keV energy range. The electron energy analyzer (e.g. cylindrical mirror analyzer) separates the emitted Auger electron by energy. An ion gun (usually argon) to sputter the specimen surface is normally available. In sputtering, a comparatively wide area of the sample surface is sputtered with the ion beam, while the primary electron beam used for specimen bombardment and Auger electron production is conducted on a much smaller locally focused point on the surface of the crater. A secondary electron detector is used to image and navigate in scanning Auger systems. The AES system also includes the associated electronics for controlling system operation and acquiring, storing, and displaying AES data.

Depending on the electron gun in the AES spectrometer, the diameter of the spot size from which analytical information is obtained can be as small as 200 Angstroms. Although some signal can be generated from backscattered electrons outside of the beam spot, the spot size is the dominant factor in the spatial resolution of AES.

Auger electrons are generated in the same interaction volume as characteristic X-rays. However, only the Auger electrons that escape the sample matrix with their characteristic Auger energy are useful for analysis. Auger electrons formed in the bulk of the sample will undergo inelastic scattering, losing a portion of their characteristic energy and becoming background signal. Depending on the material being analyzed and the energy of Auger electrons being measured, the sampling depth in AES covers the range of 20-50 Angstroms. The mean free paths and small escape depths of Auger electrons provide the surface specificity of AES.

Detection limits for most elements are on the order of 0.1-1%, in reasonably homogeneous samples. Detection limits can be affected by spectrometer operating conditions, sample surface properties such as cleanliness and roughness, signal-to-noise ratios, and the conductive character of the sample.

13.3 AES DATA COLLECTION AND ANALYSIS

As with most scanning beam microscopes and analysis tools, several characterization modes are available: spot, line scanning and mapping. The added dimension of depth profiling is afforded by ion sputtering. For AES, spot analysis is the most direct and most often used mode. In this method the primary electron beam impacts the specimen surface at the exact location of interest. Spot analysis with surface sputtering is also possible. The ion gun sputters a crater of increasing depth while spot analysis is conducted in the surface of the crater as successive layers of the specimen are exposed (see Figure 13.2). This provides a profile of specimen composition as a function of depth. Rastering the beam helps reduce charging and causes less local surface damage of the specimen.

The data for spot analysis is acquired as N(E), the intensity of electrons versus E, electron energy. Data can be displayed as full energy scan for all elements (spectra) or certain energy windows selected for particular elements of interest can be recorded. Spectra are typically displayed in one of four modes: N(E) versus E (the energy distribution of detected electrons), dN(E)/dE versus E (the differential mode), d[ExN(E)]/dE, or [ExN(E)] form. AES peaks are generally superimposed on a large energy background, so they are most easily resolved in the differential mode. However, it is often best to work with the least-processed data possible, e.g. N(E) or ExN(E) directly. This is advantageous for several reasons, including better signal-to-noise ratio and best possible energy resolution of the energy analyzer. Since measured signal is directly proportional to the number of atoms in the specimen, these forms provide the best information for quantitative analysis.

In line scanning, the primary beam is scanned repeatedly across a sample surface in one line, producing a one-dimensional map of elements present in conjunction with surface features. In mapping, the beam is scanned or rastered repeatedly over a selected area of sample surface. The electron intensity for a particular electron energy of interest is mapped as function of position. This produces a two-dimensional map of elements present in the scanned area of the specimen.

The AES energy peaks are used to identify which elements are present in the surface analyzed. Many AES peaks overlap, creating interferences and difficulties in element identification. In complex materials, a final decision about presence or absence of a given element must consider the presence or absence of the various Auger transitions that are possible for that element. Qualitative elemental identification depends on sensitivity and the signal to noise ratios observed for the element. Properties unique to a given element have a significant impact on sensitivity. These include Auger electron yield, the Auger electron energy, and ionization cross section. Characteristics of the measuring system, such as primary beam energy and current, angle of incidence of the primary beam, and the acceptance angle and energy resolution of the energy analyzer also impact sensitivity. Measurement system operating parameters can, to a certain extent, be varied to optimize signal-to-noise ratios for best elemental sensitivity and to minimize peak shifts and distortions caused by sample charging.

The number of Auger electrons obtained from a sample will, to a first approximation, be proportional to the number of atoms of that element which are present in the sample volume being analyzed. However, for quantitative analysis, sensitivity factors must be developed for each element. This is due to complications in the energetics of Auger electron emission. Since relatively few electrons escape a sample surface without energy loss, a "tail" in AES data peaks is generated which contributes to measurement background. Instrument variables that affect sensitivity factors include the primary electron beam current, and energy and the modulation energy in dN(E)/dE spectra. Sensitivity factors for atomic concentrations are applied to signals of all elements detected and are added up and divided by the total to calculate relative atomic percentages for all elements in the sample being analyzed. Sensitivity factors are unique for each sample matrix and set of analysis conditions. Sensitivity factors can also depend on the "hardware" and configuration of each AES analysis system.

In some cases, information about the overall electronic surroundings of atoms, as induced by their near-neighbor atoms, is also available from AES data. An energy shift in the position of an AES peak maximum and/or a change in shape of the peak can characterize oxidation states and atomic surroundings of elements in the sample matrix. Spectra often exhibit peak energy shifts or changes in line shape. Silicon and aluminum provide useful examples of peak energy shifts. LMM Auger transitions for silicon occur at 92eV for elemental silicon and at 76eV for silicon dioxide. Similarly, the LMM transition for aluminum at 68eV is shifted to 51eV for aluminum oxide.

Volatile hydrocarbons in ambient air are easily adsorbed onto all surfaces, and all specimens that are not "pre-sputtered" slightly to remove native carbon will appear "dirty". In addition, surface oxidation will also alter AES results as shown in Figure 13.3.

13.4 SPECIMEN, MATERIAL AND AES OPERATIONAL CONCERNS

Artifacts associated with spectrometer operating parameters and material surface properties can affect AES data. Electron beam and ion beam bombardment are energetic processes. Thus interactions with particular specimen material types under such bombardment must be considered.

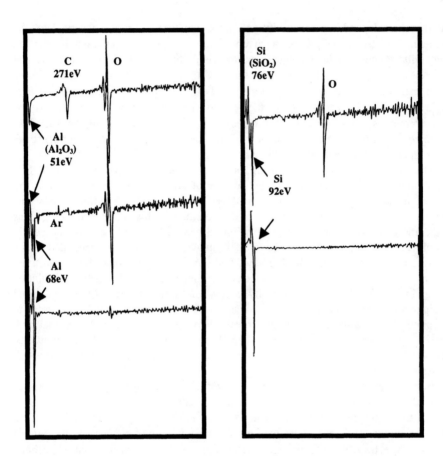

Figure 13.3 Auger spectra (dN/dE versus kinetic energy) for aluminum and silicon samples illustrate shifts with oxidation for aluminum from 68 to 51 eV (left) and silicon from 92 to 76 eV (right). As-received samples (top) show oxidation and, in the case of aluminum, organic (carbon) contamination. After sputter the oxidation and contamination is removed. Note the argon peak in the middle aluminum spectrum after sputtering of the sample.

A primary electron beam of 2nA, focused onto a 0.1 μm (1000 Angstrom, 100nm) diameter spot, produces a current density in that region of about 20 amp/cm^2. Insulating materials can not dissipate this much current and charge accumulates, often altering characteristic AES elemental peak positions and affecting the certainty of elemental identification. Surface heating can also physically alter the composition of the surface by desorbing or even melting materials. As in all beam analysis methods, insulating dielectric films in IC devices are troublesome in AES analysis because of their tendency to accumulate charge. In particular, PSG and BPSG dielectrics may desorb phosphorus or boron during AES analysis. Electron beams can be defocused to cover a larger area, reducing beam current density at the cost of spatial resolution. Beam energy and sample tilt can be experimentally optimized for particular samples to achieve the best penetration depth. Surrounding a sample with metal foil or other conductive material can help to drain charge away.

Different elements in compounds sputter under argon or other ion bombardment at different rates. An analyzed surface can become depleted of the elements with higher preferential sputtering rates, altering the specimen surface and giving false composition results. Analysis conditions must be carefully defined where compounds are being analyzed. Known reference samples should be analyzed under identical conditions to establish sputter rates.

A consequence of sputter depth profiling is the physical roughening of the sputtered crater surface. The resulting roughness factor can equal or exceed the typical escape depth of Auger electrons. This can complicate interpretation of film surface composition or interfacial information.

AES analysis requires a very clean vacuum condition, generally 10^{-9} pascal or less. High vacuum is required both for the electron spectroscopy and to maintain sample cleanliness. Due to the surface analysis nature of AES, adsorbed gases and surface oxidation can significantly alter the spectra of specimens. The adsorbed gases and surface oxidation can be sputtered away but high vacuum is required to prevent reoccurrence. Specimen materials and their surfaces must be stable and not desorb or outgas in high vacuum, otherwise AES data will not truly represent the sample's surface condition. Samples that have a significant vapor pressure of their own are difficult to analyze by AES as they can degrade vacuum levels, damage vacuum pumps, or contaminate analysis systems.

There are few absolute or standard reference materials available for AES analysis of microelectronic devices. One sample may be sufficient for qualitative identification of an easily observable anomaly. In many failure analysis applications, full characterization requires analyzing and comparing an identical area or feature of a known "good" sample to that of a known "bad" sample. This is sometimes accomplished by putting two different specimens into the spectrometer. But the "good" versus "bad" comparison can often be accomplished by analyzing two areas on the same specimen. These are often areas or features that are only a few microns apart, but which have distinctly different electrical performance or distinctly different physical appearance under microscopic examination.

13.5 AES IN FAILURE ANALYSIS

AES is particularly valuable in the failure analysis of die-related problems because of its ability to characterize extremely localized areas of contamination[7-9]. Depth profiling of interfaces within regions of device construction is also a critical application.

The value of depth profiling is clearly illustrated in an unstable STIC (Silicon Tantalum IC) resistor example[10]. The STIC resistors were 20nm thick tantalum nitride films and the failures described exhibited instability with as much as a 100% increase in resistance. The surface specificity of AES in the sputter depth profiling mode enabled a compositional comparison to stable resistors. The AES depth profiles are shown in Figure 13.4. The stable resistor in Figure 13.4a was composed of a surface-passivating "skin" of tantalum oxide (Ta_2O_5), a resistor "body" containing only tantalum and nitrogen, with signal for silicon and oxygen increasing at the interface of the tantalum nitride with substrate silicon dioxide. In contrast, the failing resistor in Fig 13.4b shows silicon and oxygen throughout the thickness of the resistor, without a discrete thickness of tantalum nitride and no tantalum oxide passivation layer. The AES data show that the unstable resistor has in effect been anodized during assembly operations.

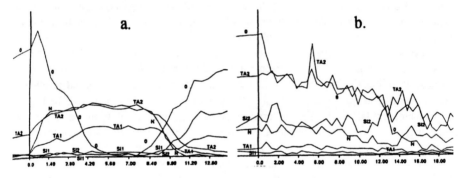

Figure 13.4. AES depth profiles (relative intensity versus sputter time) of STIC resistors, showing discrete TaN with no incorporated oxide in a good resistor (a.) and oxide incorporated in throughout the bad resistor (b.).

The use of AES for very localized analysis is illustrated by an open via analysis[11]. A device with an isolated resistive via was deprocessed through complete removal of metal-2 without overetching, preserving the vias for analysis. Microscopic and SEM inspection revealed what appeared to be a film of some kind at the interface of the failing via which was not visible on an adjacent functional via. AES analysis of via interfaces as exposed by deprocessing with no surface sputtering revealed a carbon signal on the open via four times higher than the carbon signal on the non-open via, as shown in Figure 13.5. After argon ion sputtering approximately 150 A of the surface layer and any surface-adhered species, the signal for carbon in the open via exceeded that in the non-open via by a factor of about ten. It was not until the 450 A depth point that the carbon signals were comparable in both open and non-open vias. The high carbon level was attributed to re-deposition of photoresist during resist strip.

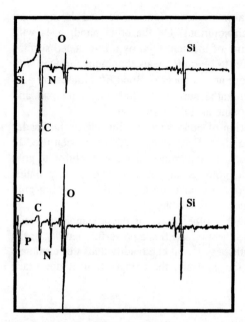

Figure 13.5. AES elemental survey of control (bottom) and failing (top) via surfaces show substantially greater carbon signal on the failing via surface.

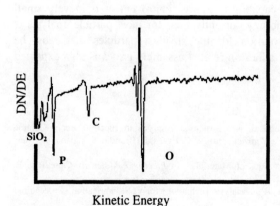

Figure 13.6. AES survey of bond pad surface shows residual silicon dioxide and phosphorus from incompletely removed passivation glass.

AES is capable of analyzing very small particles. For example, AES was used to examine micro-defects 200nm in diameter[12]. The defects appeared after plasma etch removal of a patterned silicon oxynitride layer. AES indicated the presence of titanium in addition to the silicon and oxygen detected by EDX. AES's surface specificity provides a much smaller analysis volume than EDS and better sensitivity for very small particles. The titanium-enriched SiO_2 particles were eventually found to have been introduced from process tool hardware in an earlier step.

AES has also been used extensively in diagnosing assembly-related issues. AES is a standard tool for characterizing bond pad cleanliness and defects which could affect bondability[13,14]. Relatively thin films of contamination on a bond pad can prevent reliable bonding. AES is well suited to identify such surface films and their thickness by depth profiling. Pads with incompletely removed passivation on their surfaces are a typical problem, as illustrated in the bond pad surface spectrum in Figure 13.6. The spectrum shows not only aluminum, but also a silicon peak at 76eV (indicative of SiO_2) and a peak at 120eV due to phosphorus, consistent with the phosphorus doped oxide passivation used on the device. AES spot spectra and maps can also identify silicon nodule precipitates in bond pad metallization. Another readily analyzed problem is localized areas of aluminum oxyfluoride, formed during bond pad patterning by hydrofluoric acid etch solutions. On the other hand, pads with discoloration (actually caused by scattering of incident light by a roughened surface texture to the pads) can be shown by AES to be quite clean aluminum surfaces with normal levels of native oxide, and therefore not a concern. Bond lift failures are also frequently analyzed by AES on the separated surfaces. Both the bond pad and bonding surface of the lifted bond can provide useful information[15].

The cleanliness and surface composition of bond wire attachments to the header leadframe are also critical for reliable bonding. For example, AES has been used to identify the connection between nickel surface contamination and bondability to gold surfaces[16]. Nickel can be co-plated with gold as an impurity from plating baths. Nickel underplating can also diffuses through gold grain boundaries, covering gold surfaces and oxidizing to make the surface non-bondable.

Adhesion is also an important issue for other parts of the assembly process. These include die attachment and solder adhesion. AES is a powerful method for the analysis of solder dewets and poor die attaches[17,18]. It can readily detect or eliminate issues such as silicon dioxide on backside gold and the composition of die attach areas of a header.

13.6 CONCLUSION

AES will continue to play a key role in the identification of highly localized contamination in wafer processes, particularly at interfaces. The areas of localized contamination requiring analysis are likely to decrease as feature sizes shrink. This should lead to an increased use of field emission electron guns for improved spatial resolution. One area where AES capability requires improvement is in very small particle analysis. The EDS technology currently used for most particle analysis is becoming limited by spatial resolution for the smallest particles that must be analyzed[12]. Its role in analysis of adhesion-related assembly issues is also expected to continue.

REFERENCES

1 Strausser YE. "Elemental, Chemical, and Structural Analysis in Electron Beam Column Instruments." In *Encyclopedia of Materials Characterization*, C. Evans, R. Brundle, S. Wilson, ed. New York, NY: Butterworth-Heinemann, 1992.

2 Fatemi H. "Micro and Nano Analyses of Materials." In *Electronic Materials Chemistry*, B. Pogge, ed. New York, NY: Marcel Dekker, 1996.

3 Meieran ES, Flinn PA, Carruthers, JR. Analysis Technology for VLSI Fabrication. Proceedings IEEE, July, 1987, 75(7), 908.

4 Davis LE, MacDonald N, Palmberg PW, Riach GE, Weber RE. *Handbook of Auger Electron Spectroscopy*, 2nd ed. Eden Prairie, MN: Physical Electronics Industries, 1978.

5 Burrow B, Morgan AE. Fundamentals and Applications of Auger, ESCA, SIMS, and RBS Analytical Techniques. International Reliability Physics Symposium Tutorial Notes, 1987, 4.1.

6 Rossiter TJ, Feliciano-Welpe D, Lowry RK, Schuessler P. Surface Analysis Techniques. International Reliability Physics Symposium Tutorial Notes, 1985, 5-1.1.

7   Pyle R, Kaushik V, Laberge P, Morris S, Martin L, Prack E, Hance R, Bridwell J. An Analytical Study of a Novel and New Failure Mechanism Observed in a High Density CMOS ULSI Device. Proceedings International Symposium for Testing and Failure Analysis, 1995, 129.

8   Katayama T, Mashiko Y, Mitsuhashi J, Koyama T, Tsukamoto K, Ikeda S, Nakayama A, Koyama H, Tsubouchi N. A New Failure Mechanism Related to Grain Growth in DRAMS. Proceedings International Reliability Physics Symposium, 1991, 183.

9   Tomioka H, Tanage S, Mizukami K. A New Reliability Problem Associated With Ar Sputter Cleaning of Interconnect Vias. Proceedings International Reliability Physics Symposium, 1989, 53.

10   Tse PK, Picard LJ, Swain JE, Brown RW, Gurnett CJ, Terefenko GJ. Failure Mechanisms of Thin Silicon Tantalum Integrated Circuit (STIC) Resistors on Multi Chip Modules. Proceedings International Reliability Physics Symposium, 1993, 94.

11   Le TT, Hoang HH, Michlowsky J. An Investigation of Open-Via Failures in ASIC Devices. Proceedings International Symposium for Testing and Failure Analysis, 1991, 129.

12   Childs KD, Paul DF, Clough SP. Analysis of Sub-Micron Defects on 200MM Wafers with an Auger Based Defect Review Tool. Proceedings Institute of Environmental Sciences, 1996, 147.

13   Dahlgren DA, Kingsley JR, Reidel PR.   Surface Analysis and Bond Pads. Proceedings International Symposium for Testing and Failure Analysis, 1990, 397.

14   Lowry RK, Linn JH.   Characterizing Integrated Circuit Bond Pads. Proceedings International Symposium for Testing and Failure Analysis, 1992, 165.

15   Ebel G, Hammer H.   Case History of Epoxy Contaminated Wire Bond Failures on Space Shuttle Hybrids. Proceedings International Symposium for Testing and Failure Analysis, 1992, 61.

16   Harman G. *Wire Bonding in Microelectronics*. Reston, VA: International Society for Hybrid Microelectronics, 1991, 91.

17   Hiraka S, Itabashi M.   The Influence of Selenium Deposited on Silver Plating on Adhesive Strength of Die-attachment. Proceedings International Reliability Physics Symposium, 1991, 8.

18   Walsh LB, Berry KA. Case Histories of Microelectronic Device Analyses Using Auger Electron Spectroscopy. Proceedings International Symposium for Testing and Failure Analysis, 1989, 373.

14

SECONDARY ION MASS SPECTROMETRY, SIMS

Keenan Evans
Motorola Inc. Semiconductor Products Sector

Secondary ion mass spectrometry (SIMS) is based on the ejection of charged atomic and molecular species from the surface of a solid sample when it is bombarded by a stream of heavy particles. J. J. Thomson[1] first observed this phenomenon in 1910. Later Arnot and Milligan[2] investigated the secondary ion emission resulting from positive ion bombardment. Herzog and Viehboeck[3] provided the basis of modern SIMS instrumentation, using an electron impact primary ion source in 1949. Other pioneers in the field constructed their own unique SIMS instruments[4-7]. The first commercial system derived from Herzog's work[8] was intended for the geochemical analysis of extraterrestrial material captured during the early years of outer space exploration. Since that time, SIMS has become an indispensable tool for the characterization and analysis of semiconductor components and materials. Its ability to detect all elements in the periodic table, excellent elemental sensitivity and inherent depth profiling capabilities make SIMS the appropriate choice for a number of critical semiconductor analysis needs. Dopant profiling, mobile ion monitoring, process contamination diagnosis, thin film characterization, interface analysis and surface analysis are just a few of the areas where SIMS can contribute to the root cause determination of microelectronics failures. In addition to the utilization of SIMS as a tool for the diagnosis of failures, the technique is a very powerful aid in the optimization of semiconductor processes, preventing failures. Typical applications in this preventive mode include the evaluation of the effectiveness of cleaning processes, monitoring the impact of new processing tools on wafer contamination levels and implant matching studies for technology transfer between fabrication sites.

There are various modes of SIMS operation and instrumentation depending on the nature of the primary ion source, the type of mass spectrometer used for analysis of the secondary ion species and the rate of erosion of sputtered species from the sample surface. This chapter will provide a basic understanding of the fundamental theory of SIMS[10], review the advantages and disadvantages of the various combinations of primary ion sources and mass spectrometers and illustrate the utility of SIMS with examples of specific problem solving applications. Finally, the limitations and future challenges facing the SIMS analyst will be discussed.

14.1 BASIC SIMS THEORY AND INSTRUMENTATION

The basis of SIMS is the use of a focused primary ion beam, generally in the range of 1-20 keV, to erode atoms from a selected region of a sample surface[11]. A charge exchange in the near surface environment results in the conversion of a portion of the eroded atoms to negative or positive ions. These secondary ions are extracted via an electrical potential and subsequently analyzed by a mass spectrometer. SIMS yields an analytical technique capable of elemental specificity with detection limits in the range of ppm-ppb (atomic concentration). Quantification is accomplished by comparison of relative ion yields from a sample of interest to those of ion implanted reference samples in the same matrix. Reproducibility approaching 1% can be achieved under ideal instrumental conditions. The current density of the primary ion beam can be varied over several orders of magnitude to produce sample milling rates ranging from a few Å/hr to several thousand Å/min., making the technique a powerful surface analysis, bulk analysis, thin film characterization and depth profiling tool.

14.1.1 Vacuum System and Ion Sources

As shown in Figure 14.1, there are three major components of SIMS instrumentation, the primary ion source and optics column; the secondary ion optics/mass analyzer; and the vacuum system. Ultrahigh vacuum (10E-9 to 10E-10 pascal) is critical for the optimal performance of a SIMS. The analysis chamber pressure is critical due to the high elemental sensitivity of SIMS. Background levels of the elements comprising the permanent atmospheric gases can vary significantly with the vacuum pressure. Additionally, the secondary ion yield of sample constituents may change dramatically based on the composition and partial pressures of the atmosphere surrounding the sample. Finally, the high surface sensitivity of SIMS demands an ultrahigh vacuum system so the sputtered sample species and permanent gases are not continuously redeposited on the freshly exposed sample surface.

A variety of primary ion sources are available for SIMS work. These include electron impact, duoplasmatron, surface ionization, electron beam ionization and liquid metal ion sources. The first SIMS instruments used electron impact primary ion sources with inert gases. In electron impact sources, a heated filament emits electrons, which are accelerated towards a metal grid by an electric potential. Inert gas atoms flowing across the path of the accelerated electrons are ionized and then extracted into the ion gun column to be focused by a series of electrostatic lenses. The secondary ion yield of these sources is quite low compared to that of reactive ion sources such as oxygen or cesium. To enhance positive secondary ion yields with inert gas electron impact sources, oxygen may be flushed over the sample surface. Oxygen flooding, however, degrades the analytical chamber vacuum and the sputter rate must be limited in order to give the oxygen flush gas enough time to react with the fresh sample surface.

Reactive primary ion sources are now the most commonly used in SIMS analysis. Oxygen primary ion beams can enhance the positive secondary ion yield for some elements by up to 10,000 times that possible with inert gas beams. Likewise, a cesium primary ion beam enhances negative secondary ion yields by factors of up to 10,000. A number of models have been proposed to explain the ionization processes inherent to the SIMS technique [12-14].

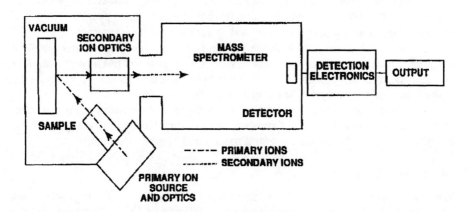

Figure 14.1. The schematic diagram illustrates the basic components of SIMS instrumentation.

The duoplasmatron configuration is used for oxygen primary ion sources. Oxygen gas is admitted into a region between an anode and cathode where a low voltage arc is sustained. This arc creates an oxygen plasma, which is dually confined by the geometric arrangement of the anode and cathode and by an electromagnetic field. A stable, high-density plasma can thus be generated and sustained in a small area. The oxygen ions are electrostatically extracted through a small opening in the anode and accelerated down the ion gun column where they are further focused and condensed by electromagnetic lenses. Ion beam currents of several mA and spot sizes of less than 1 µm can be achieved.

The surface ionization design is mainly used for cesium primary ion sources[15-17]. An ampoule of cesium metal is heated to around 120 degrees C and the cesium vapor is sent through a porous tungsten frit heated in the range of 1200 degrees C. Positive cesium ions are emitted from the surface of the tungsten frit with a very high ionization efficiency. The primary ions are similarly extracted and focused down the ion column. Current densities and spot sizes are comparable to those achieved with duoplasmatron sources.

Liquid metal ion sources can achieve the smallest primary ion beam spot size and thus the best lateral imaging resolution of the commonly used primary ion sources. Spot sizes down to 50 Å have been realized. In the liquid metal ion source, a substrate needle supports a covering of heated liquid metal, usually either gallium or cesium although other metal combinations have been used. A strong electric field is applied to electrodes in proximity to the supported liquid metal and a projection know as a Taylor cone is formed on the liquid metal surface[18]. A high intensity stream of ions is emitted from the tip of the Taylor cone[19] and focused with a series of high resolution electromagnetic lenses to a fine spot size which can be rastered over the sample surface. The small spot size achievable with liquid metal ion guns makes them ideal sources for high resolution SIMS imaging, however the elemental sensitivities for such sources are significantly below those possible with the aforementioned oxygen duoplasmatron and cesium surface ionization sources.

The ion gun columns used with the various types of primary ion sources apply a potential to accelerate primary ions out of the source region and use a combination of physical apertures and electromagnetic lenses to focus and raster the ion beam on the desired area of the sample. Manipulation of the physical apertures and focusing lenses provides control over the accelerating voltage, current density and spot size. Mass filters and curves in the columns may be incorporated into the ion gun column design to minimize the effects of neutral species and beam contaminants.

14.1.2 Mass Analyzers

The three main types of mass analyzers employed in SIMS instrumentation are quadrupole, double focusing magnetic sector and time-of flight. There are distinct advantages to each of the types of mass spectrometers with respect to cost, switching speed, secondary ion transmission, mass resolution and molecular weight ranges.

In a quadrupole mass spectrometer four parallel, electrically conducting rods are symmetrically located around a circle, which has a radius 0.86 times the radius of the rods. Potential is applied to the rods such that opposite pairs of the rods have a like potential and adjacent pairs of rods have equal but opposite potentials. The potentials applied to the rods have both a DC and RF component. An ion accelerated down the longitudinal axis of the region between the rods will oscillate down this space under the influence of the alternating RF field. Under a given set of RF and DC potential conditions ions of only a particular mass to charge ratio, m/e, will undergo a stable oscillation and be able to traverse the length of the quadrupole without striking the electrode rods and being extinguished. An ion that is successfully transmitted down the quadrupole is detected and counted by an ion multiplier. By sweeping the RF and DC fields a full mass spectrum can be acquired over the mass range covered by the spectrometer. The voltages can be changed very rapidly and thus the switching speed between different masses is very high. This means that a large number of masses may be monitored during a depth profile of a sample and that both positive and negative secondary ions may be monitored during the same depth profile acquisition. The mass range that quadrupole systems can effectively separate has an upper limit of ~ 1,000 amu. Quadrupoles also require an energy filter preceding the quadrupole section of the mass spectrometer since they are inefficient at separating masses of widely varying energies. Transmission through the quadrupole is relatively low, on the order of 1% at best and the mass resolution of quadrupoles is considerably less than the other two main types of SIMS mass analyzers. Quadrupoles are not able to separate ions of the same nominal m/e. On the other hand, quadrupole mass spectrometers are relatively inexpensive.

Double focusing electrostatic-magnetic sector mass spectrometers contain an electrostatic analyzer which provides energy separation. The ions entering the radial electrostatic sector will follow a trajectory, which depends on the energy of the ion and the strength of the electric field. Thus ions of the same energy will follow the same path. The energy selected ions then are focused through a magnetic sector where they follow a trajectory which depends on the momentum of the ion and the strength of the magnetic field. Thus ions of the same m/e will follow the same path and be able to enter the physical aperture of the detector. Double focusing mass spectrometers have a high transmission, on the order of 10-20 % and can achieve high mass resolutions of up to $m/\Delta m = 10,000$. They are more expensive than quadrupole systems and have an inherently slower switching speed. Only ions of a single polarity can be monitored during a single depth profile acquisition.

Time-of-flight (TOF) mass spectrometers are based on the principle that ions of different mass which are accelerated to the same kinetic energy, will acquire different velocities. The mass variant ions will then require different times to traverse a field free drift path. Lighter ions, which leave the acceleration region simultaneously with heavy ions, will arrive at the detector situated at the end of drift path prior to the heavier ions. TOF systems are commonly used in conjunction with a pulsed primary ion source, which has a pulse width less than 10 ns with a duty cycle on the order of several μs., Transmission for TOF systems is on the order of 10%. High mass resolution can also be achieved and TOF systems are capable of analyzing very high mass ranges. They have found widest application in organic surface analysis using an extremely low primary ion beam current density to minimize sample damage.

14.2 OPERATIONAL MODES, ARTIFACTS AND QUANTIFICATION

The two basic modes of SIMS operation are dynamic and static. Dynamic SIMS is performed with a relatively high primary ion beam current density of up to several A/cm^2. The high current density results in a rapid sputtering of the atoms from the sample surface. Dynamic SIMS depth profiling can provide sensitivity to determine ppb elemental concentrations as a function of sputter depth into the sample. Static SIMS is used to obtain information from the outermost monolayer of a sample with minimal disturbance of the composition and structure. Ion current densities on the order of a few nA/cm^2 will require several hours of bombardment to erode a single monolayer.

Survey spectra, ion imaging and depth profiling are the basic types of SIMS data acquisition used in practical application. Each of these types of data acquisition are subject to a number of SIMS artifacts that must be considered in the interpretation of the data. Secondary ion yield can vary due to bombardment angle and sample/ion beam equilibration. The secondary ion yield can also be influenced by sample composition, particularly native surface oxide (oxygen yield enhancement). The sample composition can also impact the sputter rate. Beam interaction with the surface can also produce artifacts such as surface roughening, mixing of the sample surface and redistribution of mobile ions. Charging can also occur, much as in a scanning electron microscope. Mass interference is also a major consideration. The analyst must be keenly aware of the implications of these artifacts that are inherent to the SIMS technique and make appropriate compensation in order to achieve the maximum level of reproducibility[10].

Quantification in SIMS requires a well-characterized reference standard that is as close as possible to the matrix composition of the samples of interest. With a microhomogeneous ion implanted reference standard and appropriate consideration for the various SIMS artifacts, accuracy to within 2 % of the standard value can be achieved with a similar level of relative precision. Quantification of a SIMS depth profile requires comparison to the primary component in the sample (silicon in implant characterization), and sputter rate calibration with a profilometer.

14.3 MAGNETIC SECTOR SIMS APPLICATIONS

Characterization of dopant profiles in silicon is one of the most common uses of dynamic SIMS in the semiconductor industry. Magnetic sector SIMS instruments are generally employed for this dynamic SIMS depth profiling application. This application takes advantage of magnetic sector SIMS high mass resolution capability and excellent sensitivity in depth profiling. Dopant profiling is used in implant matching studies routinely performed on blanket test wafers from ion implanters. Such monitoring procedures enable the matching of one implanter to another in terms of the implant energy, dose level and channeling characteristics. Precise configuration of SIMS samples and optimization of SIMS analysis conditions is required to achieve 2% relative quantification precision that is required for these kinds of implant matching studies[20].

SIMS is also the most powerful and versatile tool for identification of contamination associated with the ion implantation due to its unique depth profiling capability, high mass resolution and excellent sensitivity. An example of SIMS sensitivity is the detection of a ^{27}Al impurity in a ^{75}As implant profile as demonstrated in Figure 14.2. The aluminum

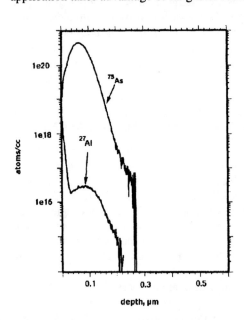

Figure 14.2. SIMS depth profile illustrates the measurement of ^{27}Al contamination in an ^{75}As Implant.

impurity dose of less than 1e12 atoms/cm² was detected in a 3e15 atoms/cm² arsenic dosage at 100 keV. The high mass resolution of magnetic sector SIMS is often required to separate species with the name nominal mass. It is worth noting that such contamination often arises from mass separation issues in the implanter, which has poorer mass resolution. For example ^{31}P has been detected in the boron implanted wafers because $^{31}P^+$ could be a mass interference for the $^{30}(BF)^+$ implant (see Figure 14.3). The high mass resolution of the magnetic sector SIMS instrument was necessary in this analysis in order to separate the $^{31}P^-$ secondary ion from the $(^{30}Si^1H)^-$ molecular ion in the silicon wafer matrix. The relationship between the contaminant distribution and the profile of nominal implanted species provides an indication of the source of the contamination.

Magnetic sector SIMS also provides an excellent method for the characterization of buried interfaces. The application of magnetic sector SIMS to oxide formation in polysilicon-silicon interface of a bipolar polysilicon emitter structure is illustrated in figure 14.4). The current gain of submicron, high-frequency npn transistors utilizing polysilicon emitter structures is influenced by a number factors including the nature of the polysilicon-silicon interface, polysilicon grain size and orientation, and dopant density[21-24]. The efficiency of pre-poly clean steps, atmosphere of the polysilicon deposition system, emitter implant and thermal profile of the polysilicon anneal process all impact the physical and chemical parameters which influence the desired electrical properties. Figure 14.4 shows depth profiles for oxygen through the polysilicon-silicon interface on emitter structures, which have been fabricated under different processing conditions. The SIMS depth profiles for oxygen are integrated for total amount of oxygen and an equivalent interfacial oxide thickness can be calculated assuming that all of the oxygen is present as a continuous silicon dioxide layer. TEM analysis confirmed the calculated oxide thickness and provided an indication of whether the thin interfacial oxide layer is a continuous layer or is intermittent in nature.

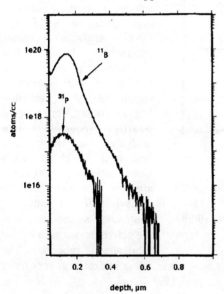

Figure 14.3. SIMS depth profile illustrates the measurement of ^{31}P contamination in a BF Implant.

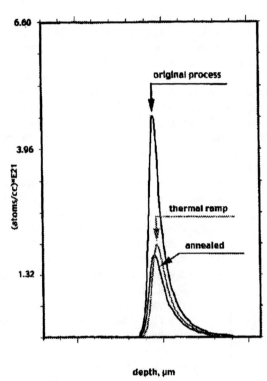

Figure 14.4. Overlay plots of the SIMS oxygen depth profiles through polysilicon/single crystal emitter interface of samples processed under three different conditions are shown.

Small area SIMS analysis on the fully processed wafer or packaged dice has always been a challenge to the SIMS community[25-26]. This is because the probe size of the favorable ion sources such as cesium and oxygen continue to lag significantly behind the aggressive shrinkage of the device feature size. In order to achieve high precision and high detection sensitivity, SIMS analysis requires relatively large test structures to provide dopant profiles for device simulation or to measure implant doses for process transfer. Frequently, failure analysis on the fully processed wafer demands more difficult small area analysis to obtain in-depth analysis of dopants or contaminants introduced into the active device area by the fabrication process. The fundamental restriction of small area SIMS analysis is a loss in detection sensitivity due to the limited number of atoms within the analytical volume[27].

14.4 QUADRUPOLE SIMS APPLICATIONS

A significant advantage of a quadrupole based SIMS system over a double focusing magnetic sector based SIMS system is the rapid switching time between species with either different mass to charge ratios (m/e) or polarities. This switching speed results in the ability to monitor a large suite of elements. This is generally required for analyses where a wide variety of potential contaminants are possible. It is also important in depth profiling where there are a large number of matrix species. Twelve or more m/e's of different polarity can be monitored in one profile with relative ease.

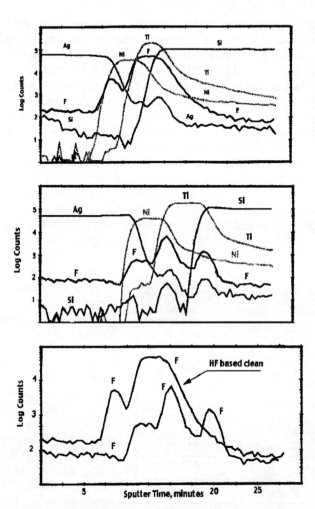

Figures 14.5-7. The profile of metallization deposited following a high F- content clean in shown in Figure 14.5 (top). The profile of metallization deposited following a low F- content premetal procedure is shown in Figure 14.6. (middle). The F distributions are overlaid in Figure 14.7 (bottom) for comparison.

The ability to monitor a large number of m/e's makes quadrupole SIMS particularly useful for the depth profiling characterization of more complex interfaces. The multilayer metallization deposited on the backside of semiconductor wafers as a solder attach is such an interface. Backmetal delamination and interfacial microcontamination are concerns for this type of die-attach. Quadrupole based SIMS is an effective method for the analysis and optimization of backmetal systems. The effect of using a high-fluoride content premetal clean instead of a low-fluoride content pre-metal procedure was studied using quadrupole SIMS to characterize the interface between the silicon substrate and the titanium/nickel/ silver metallization. Twelve mass species were monitored in each profile. The results are shown in figures 14.5 and 14.6. As shown in Figures 14.7, the difference in the fluorine content at the titanium/silicon interface was remarkable and easily detected.

14.5 TIME-OF-FLIGHT SIMS APPLICATIONS

TOF-SIMS finds its primary application in surface analysis since is commonly used with pulsed primary ion sources. Pulses sources remove material from the sample surface very slowly. Thus depth profiling is normally limited to near surface features. These features tend to make TOF-SIMS most useful for the analysis of surface contamination. Surface contamination frequently involves organic materials, which generate very complex mass spectra. This complexity arises due to the large number of fragmentation possibilities in complex organic molecules. A milky white haze on the outer surfaces of reticle boxes, observed after long-term storage in a reticle storage area[29], provides an illustrative example. FTIR analysis of haze residues on the Bayon plastic boxes identified the haze as an anti-oxidant compound, distearyl thiodiproprionate, DSTP. TOF SIMS was utilized to determine whether the DSTP could be vapor phase transported to surfaces of pellicles. Both positive and negative static TOF-SIMS spectra were collected from pellicles stored in the hazy recticle boxes. Phthalate plasticizers and siloxane compounds, which are commonly observed in wafer fab clean room environments, contributed an organic background, which has been studied[30]. Several unique peaks of much higher relative intensities were observed on the outer pellicle surface. These peaks correlated with the fragmentation pattern of a stearic acid derivative, such as DSTP. The strongest signals in the TOF-SIMS spectra were for high molecular weight secondary ion species produced by the cleavage of the stearic ester portion of the compound to form stable negative secondary ions of the general formula $C_xH_yO^-$ (See Table 14.1). This mass spectrum clearly represents the fragmentation pattern of a stearic ester compound and provides evidence that the pellicles or bare reticle surfaces were contaminated with DSTP that outgassed from the reticle boxes.

Measured Mass	Formula	Calculated Mass
127.036	$C_6H_7O_3^-$	127.039
197.20	$C_{13}H_{25}O^-$	197.191
199.21	$C_{13}H_{27}O^-$	199.206
209.15	$C_{14}H_{25}O^-$	209.191
211.22	$C_{14}H_{27}O^-$	211.206
227.25	$C_{15}H_{31}O^-$	227.237
241.27	$C_{16}H_{33}O^-$	241.253
255.31	$C_{17}H_{35}O^-$	255.269
267.33	$C_{18}H_{35}O^-$	267.269
269.33	$C_{18}H_{37}O^-$	269.284

Table 14.1. The negative secondary ions from the DSTP are shown. They are typical mass of $C_xH_yO^-$ fragments from a stearic acid derivative.

It is fairly clear from the above example that interpretation of organic mass spectra is not always simple. However, the complexity of spectra can also provide relatively conclusive identification when matches are found. It should also be noted that TOF-SIMS is equally applicable to inorganic analysis of surfaces and near surface interfaces. For example, it can provide more sensitive analysis for surface metal contamination than TXRF.

14.6 FUTURE SIMS ISSUES

SIMS has clearly become an indispensable tool in today's semiconductor industry. It is widely used for characterization as a part of failure analysis, process development, yield improvement, technology transfer, quality control and device characterization. SIMS can rapidly and routinely provide high precision measurements with sensitivity in the ppb range. There are a number of challenges facing the SIMS community as the microelectronics industry continues to shrink feature sizes. Over the course of the next several years it is predicted that depth profile resolution for SIMS instruments will be required to be less than 1 nm[31]. Commercially viable advances in primary ion source design, such as stable low energy sources[32] or non-linear extraction optics[33], will be necessary to achieve such resolution. Small area analysis needs will also increase as device dimensions shrink. As the total volume of material being removed for analysis becomes smaller, secondary ion yields and mass spectrometer transmission will have to increase in order to maintain the same level of sensitivity.

Other challenges facing SIMS include the needs to minimize the complexity of the instrumentation, reduce the cost of ownership and develop a tool that is suitable for use in a clean room wafer fab environment[31].

ACKNOWLEDGMENT

The author would like to acknowledge the contributions of Motorola colleagues Shifeng Lu, Dr. Thomas Anderson and Dr. Jane Gates.

REFERENCES

1 Thomson JJ. Philos. Mag., 20, 1910, 752.
2 Arnot FL, Milligan JC. Proceedinging Royal Society A, 1936, 156, 538.
3 Herzog RFK, Viehboeck F. Phys. Review, 1949, 76, 855.
4 Honig RE. Journal of Applied Physics, 1958, 29, 549.
5 Bradley RC. Journal of Applied Physics, 1959, 30, 1.
6 Beske HE, Agnew Z. Phys., 1962, 14, 30.
7 Werner HW. Philips Tech. Review, 1966, 27, 344.
8 Herzog RFK, Poschenrieder WP, Rüdenauer FG, Satkiewicz FG. NASA Contract No. NAS5-9254, final report, GCA-TR-67-3N (1967).
9 Liebl H, Herzog RFK. Journal of Applied Physics, 1963, 34, 2893
10 Benninghoven, A., Rüdenauer, F.G, Werner H.W. (Eds.). *Secondary Ion Mass Spectrometry.* New York: John Wiley & Sons, 1987.
11 Williams P. On Mechanisms of Sputtered Ion Emissions. Applied Surface Science, 1982, 13, 241.

12 Anderson CA. Int. J. Mass Spec. Ion Phys., 1970, 3, 413.
13 Williams P, Evans, Jr. CA. Anomalous Enhancement of Negative Sputtered Ion Emission by Oxygen. Surface Science, 1978, 78, 324.
14 Slodzian G. Some Problems Encountered in Secondary Ion Emission Applied to Elementary Analysis Surf. Sci., 48, 1975, 161.
15 Valyi, L. *Atom and Ion Sources.* London: John Wiley & Sons, 1977.
16 Storms HA, Brown KF, SteinJD. Evaluation of a Cesium Positive Ion Source for Soc. Ion Mass Spec. Anal. Chem., 1977, 49, 2023.
17 Williams P, Lewis RK, Evans, Jr. CA, Hanley PR. Evaluation of a Cesium Primary Ion Source on an Ion Microprobe Mass Spectometer. Anal. Chem., 1977, 49, 1399.
18 Taylor GI, McEwan A. J. Fluid Mech., 1965, 22, 1.
19 Gomer R. On the Mechanism of Liquid Metal Electron and Ion Sources. Applied Physics, 1979, 19, 365.
20 Lu S, Schenk R, Golonka L, Evans K. Implant/Diffusion Process Matching Characterization via Secondary Ion Mass Spec. (SIMS)" Ion Implant Technology Symposium, June, 1996.
21 Yu Z, Ricco B, Dutton R. A Comprehensive Analytical and Numerical Model of Polysilicon Emitter Contacts in Bipolar. IEEE Trans. on Elec. Dev., 1984, ED-30 (6), 773.
22 Wolstenholme G, Jorgenson N, Ashburn P, Booker G. An Investigation of the Thermal Stability of the Interfacial oxide in Poly Crystalline Silicon Emitter Bipolar Transistors by Comparing Device Results with High Resolution TEM Observations. Journal of Applied Physics, 1987, 61, 1.
23 Ning T, Isaac R. Effect of Emitter Contact on Current Gain of Silicon Bipolar Devices. IEEE Trans. on Elec. Dev., 1980, ED-27 (11), 2051.
24 Wu CY. Current Gain of the Bipolar Transistor. Journal of Applied Physics, 1980, 51, 9.
25 Dowsett, M.G. and Clark E.A. "Ion and Neutral Spectroscopy." In Practical Surface Analysis (Second Edition), Vol. 2, 229, Briggs, D. and Seah M.P., ed. London: John Wiley & Sons, 1992.
26 Lu SF, Mount GR, McIntyre NS, Fenster A. Surf. Interf. Anal. 1993, 21, 177.
27 Zeininger, H. and Criegren, R. V. In Secondary Ion Mass Spectrometry, SIMS VII, 419. Benninghoven, A., Evans, C. A., McKeegan K. D., Storms, H. A. , and Werner, H. W., eds. London: John Wiley & Sons, 1989.
28 Rasser, B., Renard, D. and Schuhmacher M. In Secondary Ion Mass Spectrometry, SIMS X, 278, Eds. Benninghoven, A., Hagenhogg, B., and Werner, H. W., eds. Chichester: John Wiley & Sons, 1997.
29 Gates J. Motorola Internal Report 960931 (1996).
30 Tamaoki M. The Effect of Airborne Contaminants in the Cleanroom for ULSI Manufacturing Process. Proceedings IEEE/SEMI Advanced Semiconductor Manufacturing Conference, 1995, 322.
31 Corcoran, S. F. In Secondary Ion Mass Spectrometry, SIMS X , 107. Benninghoven, A., Hagenhogg B., and Werner, H. W., eds. Chichester: John Wiley & Sons, 1997.
32 Smith N. S., Dowsett, M. G., McGregor, B. and Phillips, P. In Secondary Ion Mass Spectrometry, SIMS X, 361. Benninghoven, A., Hagenhogg B., and Werner, H. W., eds. Chichester: John Wiley & Sons, 1997.
33 Clegg JB. Optimum Beam Energy for High Depth Resolutions Secondary Ion Mass Spectrometry. J. Vac. Sci. Technol., 1995, A13 (1), 143.

15

FA FUTURE REQUIREMENTS

David P. Vallett
IBM Microelectronics Division

Failure analysis, like any facet of the semiconductor industry, is challenged to continuously improve to remain viable. Consider a microprocessor in the year 2006, operating at only 3 GHz, a full 500 mHz below its specified clock frequency of 3.5 gHz. Somewhere in its 200,000,000 transistors, each with a gate dielectric 1.5-2 nm thick (about 3 silicon atom spacings), buried in 7 levels and 5000 meters of metal wiring and 4000 area-array solder balls, is a single defect, perhaps an area of chemical contamination in the air-filled dielectric. It causes a slight increase in capacitance between two conductors and a circuit delay of just 2-3 ps. The contamination is invisible in an optical or electron microscope and below the detection limits of current scanning probe techniques. It is easily removed during deprocessing of the dielectric film. The IC is flip-chip mounted directly to a printed circuit board and must be exercised in-place to recreate the failure. The use of built-in diagnosability is limited and the design is highly synthesized, restricting understanding of circuit operation.

Integrated circuit technology continues to progress at an unprecedented pace that demands new failure analysis tools and techniques. Almost no area of the diagnostic process goes unchallenged by increased complexity, shrinking dimensions, lower signal margins, and smaller defects. A growing number of IC features and the defects that afflict them are already below the physical spatial resolution limit of white-light microscopy. Flip-chip packaging and numerous, dense wiring planes block access to the active area for fail site isolation and inspection tools. Subtle anomalies that were once 'cosmetic' defects are now 'killer' defects as power supply voltage drops and operating frequencies increase. Many current failure analysis methods will become quite limited in applicability while others fall into obsolescence. Failure to address these issues pro-actively will affect problem resolution cycle time, success rate, and cost. This will in turn impact time-to-market, yield, reliability, and customer satisfaction.

This chapter covers the semiconductor technology developments that drive future requirements for failure analysis. The impact of these trends on failure analysis capability will be discussed and future requirements defined. As the IC technology may evolve to a point where exhaustive analysis is impractical, alternatives to full root cause failure analysis will also be examined.

15.1 IC TECHNOLOGY TRENDS DRIVING FAILURE ANALYSIS

The National Technology Roadmap for Semiconductors highlights a number of 'grand challenges' that will demand special attention from the microelectronics community[1]. The core areas cited are scaling and lithography to enable minimum feature sizes below 100 nm; new gate, dielectric, and interconnect materials and structures; and GHz circuit speeds that will demand tight signal margins and parasitic control. Table 15.1 shows selected elements from the roadmap that particularly impact failure analysis. Although numerous changes in IC technology are forecast the three general areas of complexity, dimensions, and performance along with the new materials, design architectures, and test methodologies required to achieve them will have the most significant impact on failure analysis.

Year	1999	2006	2012
Minimum feature size (nm)	180	100	50
Complexity			
• Microprocessor transistor count (millions)	21	200	1400
• Wiring levels	6-7	7-8	9
• No. of chip IO	2000	4000	7300
Dimensions			
• Critical defect size (nm)	90	50	25
• Gate dielectric thickness (nm)	3-4	1.5-2	<1.0
• Interlevel contact diameter (nm)	200	110	60
• Junction depth (nm)	36-72	20-40	10-20
Performance			
• Logic clock speed (mHz)	1250	3500	10000
• Minimum logic supply voltage (V)	1.8-1.5	1.2-0.9	0.6-0.5
• Max. I_{off} (nA/um at 25 deg. C)	1	3	10
• Gate delay metric (CV/I) (ps)	12-13	7	3-4

Table 15.1. Selected parameters that impact failure analysis from the National Technology Roadmap for Semiconductors[1].

One pervasive shift that will have widespread effects throughout the failure analysis process is the changing nature of defects. Clearly as technology scales downward, ICs will be increasingly sensitive to smaller and subtler defects. Electrically, a defect is generally nothing more than a parasitic electrical parameter such as resistance, capacitance, or inductance that drives a circuit element to perform outside of its normal operating limits. The effect is a fault that subsequently propagates to an output. Chips operating[1] well into the GHz range with power supply voltages of 1 V or less, and transistor threshold voltages essentially fixed by short-channel effects and leakage mechanisms, will clearly have tighter margins for error. Circuits will be increasingly sensitive to lower values of unwanted capacitance and higher values of resistance or inductance. These parasitic effects will naturally take the physical form of smaller particles, thinner films, smaller holes, shorter gaps, or lower concentrations of chemical contamination.

A second major change is the move to flip-chip and related packaging styles like chip-on-chip and chip-on-board. Exacerbated by increased wiring density and fill shapes used for chemical-mechanical polishing (CMP) control, flip-chip packaging renders defects inaccessible by traditional fail site isolation methods from the front or topside of the chip.

15.2 GLOBAL FAILURE SITE ISOLATION

For memory elements like SRAM, DRAM, and cache, global fail site isolation is usually accomplished with a bit fail map created from test data. Determining the logical and physical location responsible for failure is generally independent of changes in IC process technology. Further, the limited density of interconnect wiring allows frontside execution of physical fail site isolation techniques like liquid crystal, photon emission microscopy, or charge-induced voltage alteration (CIVA). For logic products like ASICs and microprocessors however, global isolation is heavily challenged by flip-chip packages, required for high performance devices, and increasing complexity and density, forcing the development of backside techniques

It is fortunate that silicon transmits infrared (IR) radiation. This permits analysis from the backside of a die while it is flip-chip mounted. Many backside techniques however require the removal of at least some of the silicon substrate in order to reduce the effects of attenuation of IR light in highly doped substrates. Parallel polishing, surface grinding, micro-milling, and reactive ion etching processes have been introduced to make such analyses possible. Concerns remain though with insufficient optical quality of the thinned surface for some techniques, overheating of the IC during test, and fragility of the thinned silicon during die removal and handling. Laser-assisted chemical etching and FIB milling techniques will provide local thinning, although they are inefficient for full-chip or large area applications. Since these processes are critical enablers to a number of failure site isolation methods, continued development is required for higher throughput tools with improved endpoint detection and optical quality of the surface after material removal.

Thermal analysis with liquid crystal or fluorescent microthermal imaging (FMI) is limited from the backside by silicon's rapid diffusion of heat from pinpoint heat sources, some of which may be buried within the interconnect layers. Infrared thermography, although diffraction limited in spatial resolution to several microns, will continue to offer some relief for isolation of relatively high heat sources. The broadened IR response of HgCdTe and InSb arrays will assist in reaching the physical limits of the technique[2]. Precise thermal isolation lacks a tractable backside solution, although one promising method takes advantage of local changes in silicon's optical properties with changes in temperature. Development of such thermal techniques is critical as certain defects generate only excess heat and may be undetectable with light emission or other methods.

Photon emission analysis should remain a universal backside fail site isolation technique for some time. Development of detectors with higher sensitivity, particularly in the region in which silicon is transparent, is critical. The aforementioned HgCdTe and InSb detectors should provide this crucial sensitivity improvement along with integrated thermal analysis capability. CCDs with improved quantum efficiency and sensitivity will also make detection of weaker emission sites possible and require less silicon thinning.

Despite the apparent longevity of emission and hot spot techniques, two potential roadblocks are ahead. In order to detect the appropriate physical phenomena the chip must be exercised with test patterns that enable excess current flow at the defect site. For some techniques, this current must be measurable above the stand-by or quiescent current (I_{DDQ}) drawn by the whole chip. For static CMOS designs, I_{DDQ} now tends to be on the order of nanoamperes. As transistor channels shorten, sub-threshold leakage current increases considerably (despite lower supply voltage), thereby raising stand-by current. At the same time, the significantly smaller defects are anticipated to have higher resistance, reducing defect-related leakage. Lower supply voltages will further reduce the defect-related current. The result is significantly lower defect current against a higher background, seen as a lower signal-to-noise ratio by the thermal or photon emission detector.

The second limitation is that of the spatial resolution of the background image on which the defect emission is overlaid for reference in navigation. Light near 1000 nm in wavelength is the shortest that silicon will transmit. As such, spatial resolution will be diffraction limited for backside work. Products with 100 nm features, for example, will require ten times this fundamental resolution. Frontside imaging in the visible light region is similarly affected but at shorter wavelengths. This is already the case with high-density DRAMs where it is exceedingly difficult to accurately place the source of light emission at a specific cell from the topside. Although image-processing techniques will stretch resolution somewhat, it is obvious that these two powerhouse fail site isolation techniques will eventually see overall effectiveness eroded.

For frontside work, integration of light emission and thermal detection schemes with a SEM (Scanning Electron Microscope) or more likely a SPM (Scanning Probe Microscope) might provide the needed jump in resolution. This approach is unfortunately intractable for backside work. In many cases today the data provided by these techniques pinpoints the exact location of defect mechanisms. But as resolution fades, these methods will provide only coarse localization that must be followed by further isolation or probing.

Global isolation involving scanned electron or photon beams is also affected by the accessibility limitations presented earlier. CIVA, relying as it does on a scanned electron beam, is ineffective from the backside. Scanned laser techniques like LIVA and OBIC however will continue to provide valuable backside isolation capability for both open circuit and leakage mechanisms, respectively. Related techniques now in development that rely on sensing temperature-induced resistance changes (TIVA) and thermocouple effects (Seebeck effect probing) will extend this capability[3]. Like passive thermal and photon emission, these active methods will offer isolation capability for some time because of the unique and sensitive nature of the detection scheme used, though the prior limitations discussed with spatial resolution and signal-to-noise will eventually degrade their capability as well.

15.3 DEVELOPMENT IN PROBING

Probing technology is increasingly limited by accessibility and small feature sizes. Features that are already below the spatial resolution of the optical microscopes used for placement challenge direct mechanical probing. FIB tools are now routinely used to deposit metal probe pads or test points on the conductors of interest, but the process is slow and tedious. In addition, ion beam effects are invasive and will become more so as defects become subtler. Probes with smaller tip diameters can be engineered, driven partially by SPM demands, but will require improved means of placement. While SEM and FIB can be used in place of the optical microscope for probe placement, electron beam effects and more serious ion beam effects can alter the device or defect under test. Developments in probe technology and more productive, non-invasive placement methods such as those used in SPM will be required to ensure that capability for fine failure site isolation and electrical characterization is available.

Electron-beam probing for AC signals will continue to be a workhorse for frontside-accessible parts, especially DRAMs. For denser logic products however it is seriously affected by the accessibility problems cited earlier. For backside application focused ion beam holes can be milled to each point of interest[4]. The process is tedious and many nodes may be obscured even from the backside by implanted source, drain, or well regions. Special e-beam test points can be designed in at pre-determined critical nodes for characterization, but the approach is prohibitive for failure analysis that may be required at random nodes throughout the IC. Further, as device operating margins shrink, charging effects of the electron beam may alter the characteristics or load the circuit under test to the point where the measurements are no longer indicative of the original failure mode.

Two techniques are now emerging for acquisition of time-dependent signals through the backside of the chip. The first uses an IR laser to interrogate source-drain nodes or test points by sensing changes in electro-absorption and electro-refraction. The data are correlated to the applied voltage across the junction and can provide high bandwidth measurements through use of a short-pulse mode-locked laser[5]. The second method is a totally passive technique that leverages the time-dependence of light emitted from saturated transistors. The light can be time-resolved with single picosecond resolution and provides an optical waveform of switching activity that is acquired for thousands of transistors in parallel, all in one image[6]. As operating speeds increase and frequency-sensitive or AC defects begin to dominate, development of non-invasive backside probing tools is a critical need to fill the void left by electron beam probing. Tools incorporating CAD navigation and schematic links with high-speed direct-dock tester interfaces are required to maintain the productivity realized with e-beam systems.

15.4 DEPROCESSING DIFFICULTIES

The key challenges to physical and chemical analysis are the introduction of new manufacturing processes and materials, as well as the ever-shrinking physical dimensions of the structures they produce. Wet and dry chemical deprocessing will continue to be the primary global deprocessing means. Traditionally the deprocessing chemistries employed have mirrored those used in the manufacturing environment, with adjustments for isotropy, endpoint detection, and selectivity to other manufacturing and defect materials. However, the advent of new types of processing such as CMP and additive dual-damascene processing for high density interconnects requires development failure analysis deprocessing methods unique to failure analysis. This is already apparent as copper becomes the standard for high density interconnect wiring. If suitable chemical etches for copper are not found, a controllable, repeatable chip level CMP process will need to be developed in place of chemical deprocessing. Polishing techniques, which have been developed for backside thinning and wafer fab CMP may provide a starting point. It should be noted that global frontside polishing will require significant improvements in planarity across the chip compared to backside thinning. In addition, edge and corner polishing differences become a much more significant issue for individual die than for wafers.

With the resistivity advantages of copper and corresponding reduced spacing comes the need for interlevel dielectric (ILD) materials with low-dielectric constant. Currently the potential materials include fluorine-doped oxides, low-density oxides, and air-filled organics. RIE (Reactive Ion Etch) will likely remain the deprocessing method of choice. Endpoint detection and selectivity to defects will be critical as planar ILD films continue to be the norm. The development of deprocessing must become concurrent with wafer fab process development as the materials and application methods become more diverse.

In cases where fail site isolation localizes a defect accurately in three dimensions, local deprocessing techniques such as ion milling, gas-assisted FIB, and laser ablation will see increased use. Localized deprocessing reduces the chance of defects healing or being removed, but the increasingly subtle defect mechanisms anticipated will prove challenging nonetheless. Chemical enhancement of both laser and ion mill material removal will be necessary for selectivity. Improved endpoint detection will also be critical.

15.5 DEFECT INSPECTION – A TIME VS. RESOLUTION TRADEOFF

Perhaps the most ubiquitous trend in IC technology is that of shrinking physical dimensions. Historically, defects on the order of one-half to one-tenth of the minimum technology feature size have defined the spatial resolution requirements for laboratory inspection and chemical analysis equipment. For any given technology generation or node, a distribution of critical defect sizes exists. Ideally the failure analysis lab must be able to image all defects within the distribution. As dimensions shrink, the resolution required of inspection tools naturally decreases.

The approximate resolution of common inspection tools as they relate to four randomly chosen technology nodes has been overlaid onto distributions of defect sizes for those nodes as shown in figure 15.1. In the near-term, shorter ultra-violet wavelengths augmented with confocal and image-processing techniques will provide incremental resolution gains over white-light microscopy. The obvious next step is to continue with shorter wavelengths into the X-ray region. A real-time X-ray technology with rapid, three-dimensional imaging could alleviate the inspection problem. An alternative solution is available in high-resolution topographical microscopes such as SEM and SPM. While SEM has become a routine inspection tool and SPM may soon follow, there is a serious tradeoff of inspection time for resolution as their application grows.

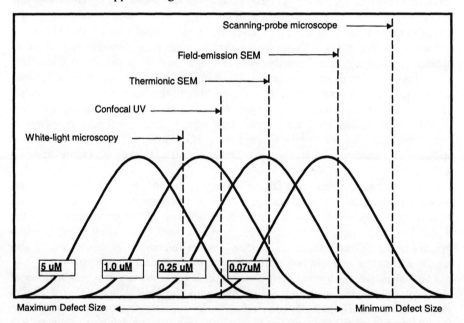

Figure 15.1. Application of inspection tools to four selected technology generations by approximate coverage of defect sizes detectable. (Courtesy of Dr. Lawrence Wagner, Texas Instruments Corp., Dallas, TX)

The reduced field of view of these tools requires more time to scan for potential defects. Images are formed based on surface or near-surface interactions, providing predominantly topographical contrast. Inspection of defects may require sample preparation to enhance contrast including removal of overlying films, possibly removing defects as discussed earlier. With SPM, the images are not formed in real-time mode and require significantly more interpretation than optical microscope or SEM images. As more samples require SEM or SPM inspection in lieu of optical microscope inspection, the factors highlighted will greatly increase failure analysis cycle-time, capital and operating costs and the human skills required.

15.6 FAILURE ANALYSIS ALTERNATIVES

Despite a strong focus on development, some areas of failure analysis will suffer physical limitations that preclude or severely limit solutions. Where FA is still possible, cycle-time or cost may become so prohibitive as to radically curtail its use. It is useful to explore reasonable alternatives to failure analysis as it is performed now.

One alternative is the use of physical and electrical signatures with a historical database to statistically classify the likely causes of failure. While signature analysis has proven quite useful on symmetrical, repetitive structures such as memory arrays, it is difficult to apply to random logic circuits. A major hurdle for logic has been the definition of signatures. Various characteristics of I_{DDQ} measurements such as pattern dependency and linearity with voltage[7] are a possibility. However, increasing sub-threshold leakages (decreasing signal-to-noise) and a greater incidence of time-dependent defects will tend to limit this approach. The possibility of numerous permutations of variables such as defect mechanism, circuit type, failure mode, process level, and process date adds further complexity and tends to detract from the statistical confidence levels. Generation of the initial data and updating the data would still require some significant level of failure analysis activity.

A shift towards more in-line failure analysis at various process levels is a natural response to complement and eventually supplement comprehensive failure analysis. This is driven by the need to shorten the feedback loops for improvements in the wafer fab and quicker time-to-yield. While useful in eliminating random manufacturing defects, design or process/product interaction problems would still require exhaustive analysis, especially for complex frequency-dependent faults.

An alternative to physical fault isolation in logic is the use of a combination of electrical failure data gathered during test with the logical and physical design elements in the CAD database – much like bit fail maps for memory devices. Design modification is required to optimize the use of such 'software diagnostics', usually in the form of scan latches or other circuits that aid in diagnosis[8]. Software tools to perform fault isolation are also required as are a design, test, and failure analysis infrastructure. While extremely powerful, the approach is not without challenges.

Flip chip is part of a trend towards chip-on-board, chip-on-chip, and stacked chip or cube. While individual die can usually be non-destructively removed from these assemblies, exercising them for either test-based or physical fault isolation will be a challenge. Software diagnosis also requires the use of fault models – logical representations of physical defects. The currently used fault models assume a node is stuck high or low. Clearly numerous other possibilities exist and models must be updated. Perhaps most acute is the problem of diagnosis resolution. Where physical fault isolation tools will often point directly to a defect, software diagnosis typically identifies one or more failing wires or nets. Follow-on physical fault isolation is required for software diagnosis, albeit with a significantly reduced range of analysis.

Development of test-based diagnostic tools with minimal design impact, based on improved fault models, and resolution comparable to a memory array is the ideal. This tool would be independent of physical constraints like packaging and shrinking dimensions. The ability to diagnose a failure directly to a defect location regardless of how deep it is buried in a chip or an assembly would clearly be a powerful failure analysis tool.

15.7 BEYOND THE ROADMAP

Current CMOS technology trends will provide more than enough challenges for failure analysis development. But there are numerous innovations that may yet be commercialized. Developments such as single electron transistors or biological computers for example would 'raise the bar' considerably for failure analysis capability. An intriguing and perhaps more imminent possibility is the spherical IC pioneered by Ball Semiconductor of Allen, TX, who has already demonstrated manufacturing processes for such devices[9]. Figure 15.2 illustrates a VLSI device consisting of a cluster of round ICs that would clearly require development of revolutionary failure analysis tools and techniques to isolate and identify defects.

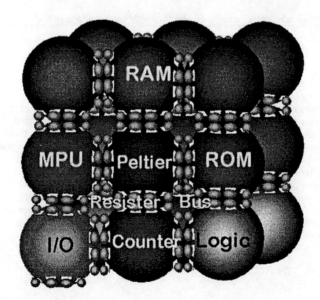

Figure 15.2. 3-dimensional VLSI cluster consisting of spherical semiconductor balls. (Courtesy of Ball Semiconductor, Allen, TX)

Regardless of semiconductor technology direction, it is clear that failure analysis development must be a pro-active, on-going effort. Industry, academia, and research community collaboration is essential to ensure that cost-efficient root cause analytical capability is available to support the accelerating IC enterprise.

REFERENCES

1 Semiconductor Industry Association. The National Technology Roadmap for Semiconductors. San Jose, Semiconductor Industry Assoication,1997.

2 Barton DL, Tangyunyong P., Soden JM, Liang AY, Low FJ, Zaplatin AN, Shivanandan K, Donohoe G. Infrared Light emission From Semiconductor Devices. Proceedings International Symposium for Testing and Failure Analysis, 1996, 9.

3 Cole EI, Tangyunyong P, Barton DL. Backside Localization of Open and Shorted IC Interconnections. Proceedings International Reliabilty Physics Symposium, 1998, 129.

4 Livengood RH, Winer P, Rao V. Application of Advanced Micromachining Techniques for the Characterization and Debug of High Performance Microprocessors. 42nd International Conference on Electron, Ion, and Photon Beam Technology and Nanofabrication , May 1998,.

5 Paniccia M, Rao V, Mun Yee W. Optical Probing of Flip Chip Packaged Microprocessors. The 42nd International Conference on Electron, Ion, and Photon Beam Technology and Nanofabrication , May 1998,.

6 Kash JA, Tsang JC, Knebel DR, Vallett DP. Non-invasive Backside Failure Analysis of Integrated Circuits by Time-dependent Light Emission: Picosecond Imaging Circuit Analysis. Proceedings International Symposium for Testing and Failure Analysis, 1998.

7 Gattiker AE, Maly W. Current Signatures: Application. Proceedings International Test Conference, 1997, 156.

8 Vallett DP, Soden JM. Finding Fault With Deep Sub-micron Ics. IEEE Spectrum, October 1997,39.

9 Ball Semiconductor Inc. web site at: http://www.ballsemi.com/default.htm

FURTHER SUGGESTED READING

- E. B. Eichelberger, et al, *Structured Logic Testing*, Prentice-Hall, Englewood Cliffs, NJ, 1991
- *IEEE Design and Test of Computers*, July-September 1997 Issue, pp. 59-97
- S. M. Kudva, et al, *The Sematech Failure Analysis Roadmap*, Proceedings of the 21st International Symposium for Testing and Failure Analysis, 1995, ASM International, Materials Park, OH, pp. 1-5
- R. E. Anderson, et al, *Future Technology Challenges for Failure Analysis*, Proceedings of the 21st International Symposium for Testing and Failure Analysis, 1995, ASM International, Materials Park, OH, pp. 1-5
- *Failure Analysis in a Nanometer World*, The Industrial Physicist, June, 1998
- M. Bahrami, et al, *Future Failure Analysis Needs*, available from SEMATECH, Austin, TX
- *Texas Instruments Technical Journal*, Vol. 14, No. 5, October-December, 1997, pp. 3-122
- D. Jensen, et al, *Assessing Future Technology Requirements for Rapid Isolation and Sourcing of Faults*, Micro Magazine, July-August, 1998
  http://www.micromagazine.com/archive/98/07/jensen.html.

INDEX

Abrasives	160-162
Analog Devices	13-14, 16, 34-38, 39
Assembly	8, 201, 225-226
Atomic Absorbtion Spectrometry (AAS)	202
Atomic Force Microscopy (also see Scanning Probe Microscopy)	189-192
Auger Electron Spectroscopy (AES)	184, 199, 201-203, 205, 217-226
Automated Test Equipment (ATE)	16-17
Automatic Test Program Generation (ATPG)	29
Backscattered Electrons	185-186
Barrel Etcher	151
Bench Test	16
Blackbody Radiation	67-70
Brightfield Optical Microscopy	179
Built In Self Test (BIST)	39
Capacitive Coupling	120-121
Charge Coupled Device (CCD)	81-82, 88, 93-94
Charge Induced Voltage Alteration (CIVA)	88, 103, 105-107, 110-111, 244
Charging	119-121
Chemical Analysis	5
Cleaving	160, 162-164
Computer Aided Design (CAD) Navigation	111, 128-131, 166, 245, 248
Confocal Microscopy	181-182
Continuity Test	17
Cross Section	5, 159
Current Sources	37
Curve Tracer	3, 15, 18-24
Customer Return	9
Darkfield Optical Microscopy	179-180
Data Converters	38
Dead Time	207, 214
Decapsulation	3, 61-64
Decoration, see Staining	

INDEX

Delidding	59-61
Delineation, see Staining	
Deprocessing	5. 145-156, 159
Depth Profiling	219, 222-226, 229, 233-236
Design Debug	6, 16
Design for Test	29, 39
Diffraction Limited Resolution	45, 47-49, 178, 182-183
Diode Characteristics	18-20
Dotmap	220
Dynamic Random Access Memory (DRAM)	13-14, 31-34, 243-245
Electrical Characterization	3, 13-39
Electrical Isolation	1
Electrical Failure Analysis (EFA)	31
Electron Beam Induced Current (EBIC)	87, 96, 102-103, 110, 113-114
Electron Beam (e-beam) Probe	113-114, 117
Encapsulation	160
Energy Dispersive Spectroscopy (EDS)	184, 189, 198-199, 203, 205-214, 218
Field Emission Scanning Electron Microscope (FESEM, also see SEM)	
	159, 166-167, 183, 196-197
Flip Chip	4, 52-54, 65-66, 113, 137-142, 166-168, 188-189, 241, 248
Fluorescent Microthermal Imaging (FMI)	67, 76-85, 243
Focused Ion Beam (FIB)	115-116, 131-139, 145, 160, 171-173, 205, 245-246
Four Point Probing	117
Fourier Transfer Infrared Spectroscopy (FTIR)	197, 199-201, 203
Functional Patterns	29-31
Grinding	63-66, 160-162
Heavy Ion Backscattering (HIBS)	198
I-V Curve, see Curve Tracer	
I/O Characterization	23-24
IDDQ	14, 17, 25-28, 34, 39, 244
Inductively Coupled Plasma (ICP)	154
Infrared Light (IR)	67-70, 84-85, 138-142, 182-183, 243, 245
Infrared Modulation-Based Probing	139-142
Infrared Thermography	67-70, 84-85
Interference Contrast	180-181
Ion Sputtering	219, 222-223
Jet Etch Decapsulation	3, 61-62

INDEX

Laser Ablation	113, 246
Laser Etching	145
Laser Decapsulation	63
Laser Scanning Microscope (LSM)	95, 181-182
Light Induced Voltage Alteration (LIVA)	87, 96-99, 110, 244
Linescan	220
Liquid Crystal	67, 70-76, 84-85
Local Field Effects	121
Logic Analysis	13-14, 15-16, 25-31, 39
Logic State Mapping	99
Low Energy Charge Induced Voltage Alteration (LECIVA)	103, 106-111
Magnetically Enhanced Reactive Ion Etching (MERIE)	154
Magnet Sector SIMS	234-236
Mechanical Polishing	5, 155-156
Mechanical Probing	114-117
Memory Devices	13-14, 31-34
Micro-Raman Spectroscopy	201
Microscopy, General	175-177
Microscopy FTIR, see Fourier Transfer Infrared Spectroscopy	
Navigation (see CAD Navigation)	
Near Field Microscopy	201
Near Infrared (NIR)	88, 91-94
Op-amps	37-38
Optical Beam Induced Current (OBIC)	87, 95-96, 99, 110, 113-114
Optical Beam Induced Resistance Change (OBIRCH)	87, 99-101, 110
Optical Microscopy	175, 177-183
Parallel Plate Etcher	151-152
Parallel Polishing	155-156
Parametric Analysis	7, 14, 15, 17, 20-23
Photon Emission Microscopy (PEM)	4, 87-94, 244
Picosecond Integrated Circuit Analysis (PICA)	141-142
Plasma Deprocessing	148, 150-154
Polishing	5, 65-66, 155-156, 160-162, 164-165
Probe Point Creation	134-135
Process Development	6-7
Quadrupole SIMS	236-237
Qualification	9

Radiancy, see Blackbody Radiation
Reactive Ion Etching (RIE) 152-154, 246
Real-Time X-ray Radiography 43, 47-55, 247
RIE Grass 153-154
Residual Gas Analysis (RGA) 202
Repackaging 64
Resistive Contrast Imaging (RCI) 87, 103-104
Root Cause 2-3

Scan 3, 29-30, 39
Scanning Acoustic Microscope (SAM) 4,5, 43-55
Scanning Electron Microscopy (SEM)
    5, 115-116, 117-120, 124, 175, 183-188, 205, 207-208, 244, 245, 247
Scanning Probe Microscopy (SPM) 175, 189-192, 244, 245, 247
Scanning Tunneling Microscopy (STM) 189-190
Scanning Transmission Electron Microscopy (STEM) 189
Secondary Ion Mass Spectrometry (SIMS) 197, 199, 203, 229-239
Seebeck Effect Imaging (SEI) 87, 99, 101-102, 110, 244
Shmoo Plot 14, 16, 30-31, 33
Signature Analysis 248
SiLi Detector 207-209
Staining, 165-166
Static Random Access Memory (SRAM) 243
Stroboscopic Imaging 126-128

Time-of-Flight SIMS 238-239
Thermal Gravimetric Testing /Thermal Mechanical Testing 203-204
Thermally Induced Voltage Alteration (TIVA) 87, 99-101, 110, 244
Thermomechanical Decapsulation 63-64
Total Reflection X-ray Fluorescence (TRXRF) 201, 208
Transistor Characteristics 20-23
Transmission Electron Microscope (TEM) 159. 168-173, 175, 189, 192, 205

Ultraviolet Light 76-80, 182-183, 192
Ultraviolet Film Bleaching 84-85

Voltage Reference/Regulator 35-36

Walkout 20
Wavelength Dispersive Spectroscopy (WDS) 207
Wedge Polishing 169-171
Wet Chemical Etch, also see Staining 148-150

INDEX

X-ray	4, 43, 47-55
X-ray, Characteristic	
X-ray Fluorescence	202
X-ray Photoelectron Spectroscopy (XPS)	201
Yield Analysis	6-8